T0276206

Gas-Turbine Power Generation

The Power Generation Series

Paul Breeze — Coal-Fired Generation, ISBN 13: 9780128040065

Paul Breeze — Gas-Turbine Fired Generation, ISBN 13: 9780128040058

Paul Breeze — Solar Power Generation, ISBN 13: 9780128040041

Paul Breeze — Wind Power Generation, ISBN 13: 9780128040386

Paul Breeze — Fuel Cells, ISBN 13: 9780081010396

Paul Breeze — Energy from Waste, ISBN 13: 9780081010426

Paul Breeze — Nuclear Power, ISBN 13: 9780081010433

Paul Breeze — Electricity Generation and the Environment, ISBN 13: 9780081010440

Gas-Turbine Power Generation

Paul Breeze

AMSTERDAM • BOSTON • HEIDELBERG • LONDON
NEW YORK • OXFORD • PARIS • SAN DIEGO
SAN FRANCISCO • SINGAPORE • SYDNEY • TOKYO

Academic Press is an imprint of Elsevier

Academic Press is an imprint of Elsevier
125 London Wall, London EC2Y 5AS, UK
525 B Street, Suite 1800, San Diego, CA 92101-4495, USA
50 Hampshire Street, 5th Floor, Cambridge, MA 02139, USA
The Boulevard, Langford Lane, Kidlington, Oxford OX5 1GB, UK

ISBN: 978-0-12-804005-8

British Library Cataloguing-in-Publication Data
A catalogue record for this book is available from the British Library

Library of Congress Cataloging-in-Publication Data
A catalog record for this book is available from the Library of Congress

For Information on all Academic Press publications
visit our website at http://store.elsevier.com/

CONTENTS

CHAPTER 1

An Introduction to Gas-Fired Power Generation

Natural gas is the second most important fossil fuel for generating electricity after coal. Its exploitation as an energy source for electrical power generation dates from the early part of the 20th century. However it was in the latter part of the century that capacity accelerated alongside the development of efficient gas turbine-based power stations. The use of natural gas for power generation has continued to grow into the 21st century, particularly among the developed nations. In the United States, where the economic extraction of natural gas from shale rocks in the past decade has dramatically increased the amount of natural gas available and reduced the cost of the gas, this has been especially noticeable. Other regions are now trying to exploit the potential for shale gas, with varying success.

While cost has been one factor driving the use of natural gas for power generation, another has been climate change. Combustion of natural gas produces significantly less carbon dioxide for each unit of electricity generated than would be emitted by a coal-fired power station during production of the same unit of power. A natural gas-fired power station can be as large as a coal-fired power station and it can generate electricity continuously like a coal-fired plant,[1] making it a good substitute for the latter. Switching from coal to natural gas has therefore provided a simple means for companies to meet climate-change emission control limits more easily and for nations to achieve targets for carbon dioxide emissions.

While a switch from coal to natural gas will reduce overall emissions for each unit of power it will not eliminate them. To achieve that requires either abandoning the combustion of fossil fuel for power generation or the introduction of carbon capture and storage technologies to natural gas-fired power plants. The technologies for both carbon capture and its storage already exist but they have yet to be tested the

---

[1]Many types of renewable power plant cannot produce power continuously as they rely on an intermittent source of energy such as the sun or the wind.

Gas-Turbine Power Generation. DOI: http://dx.doi.org/10.1016/B978-0-12-804005-8.00001-X

scale of a large natural gas-fired power station. So, while power companies and some governments are promoting natural gas—with or without carbon capture—as a stage on the way to a carbon free energy economy, environmental campaigners generally consider it to be a distraction, sapping investment that could otherwise be used to develop more extensive renewable generation resources.

The recognition of natural gas as a source of energy can be traced back to around 500 BC. Natural gas seeping from the ground will often ignite to create a flame but in earlier times its significance as an energy source was not recognized. More often the flame was instead considered a sacred sign. It was in China for the first time that this seeping gas was collected and transported in bamboo pipes, then used to boil sea water and produce salt.

Commercial use of gas for energy originated in Great Britain in the 18th century although this was based on gas manufactured from coal rather than natural gas extracted directly from the earth. The modern natural gas industry probably started in the United States in the middle of the 19th century when gas was first pumped from a natural gas well and distributed for lighting. At this stage, however, long distance transportation of gas was not possible and gas was all used locally. It took the development of gas pipeline technology in the 20th century to make natural gas widely available.

Pipelines are a key technology in the growth of the natural gas industry. The cheap, easy transportation of natural gas though pipelines gives the fuel one of its main attractions. Pipeline delivery has allowed gas to be used as a source of domestic energy for both heating and cooking, as a commercial energy source, and in industrial processes. This ease of transportation has also allowed natural gas to be traded globally, like oil, and this has led to large global and regional fluctuations in the cost of natural gas over the past 30 years. Such fluctuations can have serious economic consequences, particularly for power plants that use the fuel.

The commercial electricity industry began at the end of the 18th century when electric power started to replace gas as a source of light, incidentally removing one of the main uses of gas at the time. This prompted gas companies to search for other uses for their product. One of those was electricity generation. The production of electricity from natural gas began during the first half of the 20th century with the use of gas in combustion boilers similar to those used for the combustion of coal. While this practice continues in some parts of the world, particularly where natural gas is abundant, it was the development of gas turbines for power generation that led to a dramatic rise in the global use of natural gas by the power industry. Gas turbine-based power plants, particularly combined cycle plants that add a steam turbine to exploit the waste heat from gas turbine exhausts, have much higher efficiencies than traditional boiler plants and are both cheap and quick to build. This has made then extremely attractive commercially. Piston engines fired with natural gas are also popular for smaller capacity generation.

Large natural gas-fired power stations have become an important component of the electricity supply industries in many parts of the world, particularly in the developed nations of Europe and in North America. However the arrival of renewable energy has changed the complexion of the electricity industry and with it the function of these gas-fired power stations. Their use is now being adapted to help accommodate the large volumes of renewable electricity generation that is being introduced into national grids in many parts of the world, with the natural gas plants providing back up for the intermittent renewable supplies. This renewable support duty is likely to provide a role for natural gas-fired power plants well into the 21st century.

1.1 THE HISTORY OF NATURAL GAS AS AN ENERGY SOURCE AND FOR POWER GENERATION

Natural gas, the product of the fossilization of organic material in the earth's crust, has made its presence known for millennia when gas leaking from reservoirs close the earth's surface has ignited to create a flame bursting from the earth. The temple to the Oracle of Delphi, an ancient Greek temple on Mount Parnassus, is reputed to have been founded after a goat herd found a flame issuing from a fissure in the rock, a flame that was considered to be divine. Similar divine flames have been identified from ancient Persian and Indian religions and are probably the original source of "eternal flames" that still have symbolic meaning today in both religious and secular contexts.

It was in China that the nature and utility of natural gas was first recognized as a source of energy. Having realized that the gas

emerging from the ground was a useful source of heat, Chinese crafts-
men collected the gas and then transported it using simple bamboo
pipelines to sites where the heat could be exploited. The best documen-
ted example is the use of natural gas to produce sea salt from salt
water by using the heat to evaporate the water. Simple use of the gas
in this way is not recorded elsewhere but French explorers in North
America in the early 17th century reported that native Americans were
igniting gas that seeped into and around Lake Erie.

The commercial gas industry from which the natural gas industry
evolved began with the large scale gasification of coal in England
around 1785 to produce "illuminating gas." This process involved
heating bituminous coal in a vessel in the absence of air, driving off a
combustible gas and leaving a solid residue, coke, which was used in
the iron industry. The coal gas produced by this method was used for
street lighting and eventually for domestic lighting. The gas production
and distribution industry developed on the back of this process which
continued to be used for gas production well into the middle of the
20th century in the United Kingdom and elsewhere. In the United
Kingdom the gas, often called "town gas," was finally replaced with
natural gas from the North Sea when exploitation of these reserves
began in the second half of the 20th century, stimulating the construction of a UK national gas grid.

The use of coal gas in the United States began in 1816 when the gas
was used for street lighting in Baltimore. It was in the United States
later in the century that the true natural gas industry began when gas
was raised from wells and used to provide lighting. Meanwhile in 1855
the German scientist Robert Bunsen developed a stable and safe means
of burning natural gas efficiently. His invention was an improvement
on earlier designs and is now known widely as the Bunsen Burner. The
burner design allowed the exploitation of natural gas as a source of
heat to develop and expand.

Natural gas could still only be used locally because of the inability
to transport it over long distances. Where it was produced in conjunction with oil[2] it was often burned (flared) at the wellhead, a practice
that continues today in some parts of the world. Some progress
was made with pipeline technology during the early decades of the

---

[2]Associated gas.

20th century and in the United States the first interstate pipelines were built. However it was the development of new metallurgical and iron-working techniques after the Second World War that made the construction of extensive modern pipeline networks possible.

The exploitation of natural gas for power generation began with the use of gas in boilers to raise steam for a steam turbine. Where natural gas was readily available this offered a cheap and reliable means of generating electricity, particularly in oil producing countries that had large volumes of natural gas available for which there was no obvious use. Such boilers are similar to coal-fired boilers but without the necessity for coal handling and preparation. While this offered a viable option where it was available, coal was the fuel of choice for most large fossil-fuel power stations.[3]

It was during the 1930s and 1940s that gas turbines, which were being developed as aero engines, began to be used in stationary applications such as electricity generation too. The first commercial gas turbine was installed in 1949 in the United States and the technology became established during the 1950s. Until the early 1980s the most common use of these "aeroderivative" gas turbines was for providing power during peak demand periods. However, the development of heavy duty gas turbines, and of the combined cycle power plant, led to a massive growth in this type of power station during the 1980s and 1990s, growth that has continued into the second decade of the 21st century. It is these combined cycle power stations that have promoted natural gas to become the second most important fossil fuel for power generation after coal.

1.2 GLOBAL ELECTRICITY PRODUCTION FROM NATURAL GAS

Until the 1970s, most natural gas used for power generation was burned in steam-raising boilers. The gas, usually a by-product of oil production, was cheap and burning it to raise steam was an economic alternative to flaring which was widely practiced. Figures detailing the amount of electricity generated from natural gas, globally, are not available for the first half of the 20th century. However, data from the International Energy Agency (IEA) show that in 1973 there were

[3]Some oil-rich countries also used oil to fire boilers for power generation although this practice declined steeply during the global oil crisis of the 1970s.

Table 1.1 Annual Electricity Production from Natural Gas

Year	Annual Production of Electricity From Natural Gas (TWh)	Total Global Annual Electricity Production (TWh)	Natural Gas Production as a Proportion of Annual Global Electricity Generation (%)
1973	740	6117	12.1
2004	3420	17,450	19.6
2005	3623	18,239	19.7
2006	3805	18,930	20.1
2007	4132	19,771	20.9
2008	4299	20,181	21.3
2009	4292	20,055	21.4
2010	4758	21,431	22.2
2011	4846	22,126	21.9
2012	5100	22,668	22.5

Source: *International Energy Agency.*

740 TWh of electricity generated from natural gas-fired power stations. That represented 12.1% of total global electricity production during the year as shown in Table 1.1.[4]

The use of natural gas for electricity generation began to accelerate during the 1980s, as outlined above, and by 2004, the total production from natural gas had risen to 3420 TWh. This now represented 19.6% of total world electricity production (Table 1.1). Since then the consumption of natural gas for power generation has continued to rise steadily both in absolute terms and in terms of the proportion of power generated from gas. In 2006 natural gas provided over 20% of total global electricity production and in 2012 the proportion was 22.5%, when total production from natural gas was 5100 TWh.

Nationally, the largest user of natural gas for electricity generation is the United States which in 2012 produced 1265 TWh from the fuel according to figures from the IEA shown in Table 1.2.[5] This is far more that the second largest user, the Russian Federation, which generated 525 TWh from natural gas in 2012, while the third largest, Japan produced 397 TWh. Other big users in the table include the Islamic Republic of Iran, Egypt, Mexico, Saudi Arabia and Thailand, all of which are major producers of natural gas.

[4]Key World Energy Statistics 2006–2014.
[5]Key World Energy Statistics 2014.

Table 1.2 National Production of Electricity from Natural Gas, Top 10 Leading Nations in 2012

Country	Electricity Production From Natural Gas (TWh)
United States	1265
Russian Federation	525
Japan	397
Islamic Republic of Iran	170
Mexico	151
Italy	129
Egypt	125
Saudi Arabia	121
Thailand	117
Korea	112
Rest of the world	1988
Source: *International Energy Agency.*	

The growth in the use of natural gas for power generation in combined cycle power plants can be seen from the rapid rise in the consumption of gas by utilities in the United States over the past two decades. In 1993 total US natural gas electricity production was 415 TWh according to the US Energy Information Administration (US EIA) and generating capacity based on natural gas was 66 GW. By 2004 production was 709 TWh while the installed natural gas-fired generating capacity was 224 GW. As shown in Table 1.2, in 2012 the total production was 1265 TWh, three times the production in 1993. The US natural gas-fired generating capacity in 2011 was 415 GW.

CHAPTER 2

The Natural Gas Resource

Natural gas, like oil, is the product of the fossilization of microscopic plants and animals that lived in the world's oceans during the Carboniferous period which occurred between 300 and 360 million years ago. The plants, probably similar to algae, flourished in the oceans where they absorbed sunlight and used the energy to fix carbon from the atmosphere. When they died, they formed layers on the bottom of the oceans. These layers eventually turned into a polymeric material called kerogen. The kerogen was usually mixed with sand, clay and minerals to form strata which as they thickened and became under greater and greater pressure, eventually forming sedimentary rocks with the kerogen inside them.

As a result of the high temperatures and massive pressures that the polymeric material experienced as it was buried deeper in the earth, over time the kerogen turned into oil or gas, the precise product depending on the form of kerogen and the conditions it experienced. The higher the temperature to which it was exposed, the lighter the oil that was eventually produced. For predominantly plant-based kerogen and the highest temperatures the result was often natural gas. The oil or gas, once produced, was gradually absorbed into the pores of the rock surrounding it.

Over geological timescales the rocks containing the oil or gas moved and deformed, often becoming folded, and might be buried beneath other layers of rock. The fate of the oil and gas within its pores depended on what happened to the rock that contained them. If the porous rock was close to the surface the oil and gas might eventually migrate to the surface and either escape in the case of gas or form visible pools in the case of oil. In other cases the migration of the oil and gas was halted by an overlaying cap of impermeable rock through which it could not pass. Under these conditions the gas and oil accumulated in the rock below the cap, creating a reservoir. These reservoirs are the main source of oil and gas supplies today.

Gas-Turbine Power Generation. DOI: http://dx.doi.org/10.1016/B978-0-12-804005-8.00002-1

The modern oil and gas industry exploits three different geological sources of oil and gas, as shown in Fig. 2.1. The fossil fuel trapped in reservoirs in sedimentary rocks below impermeable caps as outlined above are the traditional (or conventional) source. In addition some oil and gas is trapped within impermeable rocks from which it cannot easily migrate. These are called unconventional hydrocarbons and include shale oil and shale gas deposits as well as "tight gas" which is locked inside sandstone. Natural gas can also be trapped within coal seams. Coal-bed methane is considered another unconventional hydrocarbon source. The gas can be recovered from coal seams that are too deep to mine by using unconventional gas extraction techniques.

From an industry perspective, the main difference between conventional and unconventional hydrocarbons is their ease of extraction. Drilling into a conventional hydrocarbon reservoir will normally release a flow of gas or oil naturally. However in an unconventional deposit the gas must be stimulated using a technique such as hydraulic fracturing (fracking). Natural gas can either be found in association with oil in a conventional reservoir or alone in what is known as a conventional nonassociated gas reservoir.

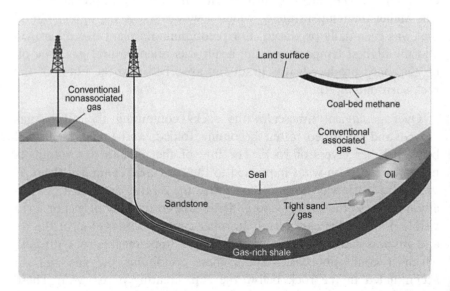

Figure 2.1 The geology of different natural gas resources. Source: From Wikimedia. US Energy Information Administration.

Natural gas can also be found in one further form, natural gas hydrates. These are a crystalline form of natural gas that is created in the presence of water, under high pressure and relatively low temperatures. The hydrates are composed primarily of methane and water although other components can also be found. Natural gas hydrates are usually located within two types of geological formation, marine shelf sediments and onshore polar regions beneath permafrost. The gas from these hydrates is relatively expensive to extract and is not exploited today. However they may represent the largest global reserves of natural gas.

2.1 THE COMPOSITION OF NATURAL GAS

Natural gas as it is taken from the ground will contain a mixture of components, the most significant of which is methane.[1] The gas usually contains between 70% and 90% methane as shown in Table 2.1. The remainder of the gas will be a mixture of higher hydrocarbons such as ethane, propane and butane—each of which may individually account for up to 20% of the natural gas—along with carbon dioxide, hydrogen sulfide, nitrogen and oxygen and traces of rare gases such as argon and krypton. In addition there is normally some water vapor present in the gas that emerges from the well.

Table 2.1 Chemical Composition of Natural Gas	
Chemical Component	Proportion in Natural Gas (%)
Methane	70–90
Ethane	0–20
Propane	0–20
Butane	0–20
Carbon dioxide	0–8
Nitrogen	0–5
Hydrogen sulfide	0–5
Oxygen	0–0.2
Source: Gail.	

[1]Sometimes natural gas comes with an associated natural gas condensate, sometimes simply called condensate. This condensate is composed of higher hydrocarbons that are liquid under normal atmospheric temperatures and pressures. Other hydrocarbons may be condensed out using specialized equipment to leave virtually pure methane.

Once it has been extracted from the reservoir, the raw gas is treated to clean it and to extract higher hydrocarbons so that the residue is primarily methane. Hydrogen sulfide, which can be processed to produce sulfur,[2] is also removed as is carbon dioxide.

The first processing stage will depend on the type of natural gas. When oil is extracted from associated wells the natural gas may be dissolved in it. In this case it must first be released and then separated from the oil. The separation will often occur naturally when the high pressure to which the mixture was subjected underground is reduced, when the gas comes out of solution leaving the oil free of gas. However there are cases in which specialized separation equipment is required, usually when the gas is dissolved in a light crude oil or in natural gas condensate. Separation, where necessary, is followed by a process to remove any water that is associated with the gas. Liquid water can be separated from the gas simply but some water may be in solution in the natural gas. This is removed either by an absorption or an adsorption technique.

The next stage is to remove the higher hydrocarbons. These, once separated, are often called natural gas liquids. Again there are two processes in common use, an absorption technique and the cryogenic expander process. The latter is more effective at separating lighter hydrocarbons, principally ethane, from the natural gas. Once the natural gas liquids have been separated then the liquids are themselves processed in order to isolate the different hydrocarbons they contain. These will usually include ethane, propane, butane and pentane. Separation may also result in a liquid called natural gasoline. All of these can be sold for use in oil refineries, in chemical processes, and as fuels.

After this hydrocarbon separation the natural gas is treated to remove carbon dioxide and hydrogen sulfide from the remaining gas stream. The amount of hydrogen sulfide extracted can be significant. In the United States, for example, this sulfur accounts for around 15% of total national sulfur production.

The result of all these processes is to leave a gas that is composed mainly of methane. A typical pipeline gas sold by a company in North America contains 95% methane, around 3% ethane, 1% nitrogen and

---

[2]Natural gas with a large component of hydrogen sulfide is called sour gas. When the amount is low it is called sweet gas.

0.5% carbon dioxide[3] with traces of oxygen and other hydrocarbons. However the exact composition will vary depending upon the source and this has to be taken into account by power companies because the energy content of the natural gas will change with composition and this will affect the operation of a power plant. Typically the gross heating value of natural gas is 36–42 MJ/m^3.

2.2 NATURAL GAS RESERVES

The world's natural gas reserves are distributed across the globe, but not evenly so. The greatest conventional natural gas reserves are found in Russia and the Middle East with significant reserves in other parts of the world. These conventional reserves provide most of the world's natural gas today. Unconventional reserves are spread more widely. According to the International Energy Agency (IEA) the conventional and unconventional recoverable resources identified across the globe are of similar size.[4] Unconventional gas production accounts for up to 60% of gas production in the United States but it contributes very little in other parts of the world. The IEA suggests that total recoverable reserves could provide natural gas for 250 years at the current rate of consumption. However BP figures below in Table 2.2[5] suggest a rather shorter lifespan for total proved reserves.

Table 2.2 Global Proven Gas Reserves by Region, 2014			
Region	Total Proved Reserves (Trillion m^3)	Regional Reserves as a Proportion of Global Total (%)	Reserve/Production Ratio (Years)
North America	12.1	6.5	12.8
Central and South America	7.7	4.1	43.8
Europe and Eurasia	58.0	31.0	57.9
Middle East	79.8	42.7	>100 years
Africa	14.2	7.6	69.6
Asia Pacific	15.3	8.2	28.7
World	187.1	100.0	54.1
Source: BP.			

[3]Figures are from Union Gas.
[4]The Golden Age of Gas, International Energy Agency, 2011.
[5]BP Statistical Review of World Energy June 2015.

According to the figures in the table, total proved reserves in the Middle East are 79.8 trillion m³, 42.7% of the global total. Reserves in Europe are 58.0 trillion m³, 31% of the total. Between them, these two regions account for close to 74% of proved reserves. Elsewhere, proved reserves are significantly lower. The Asia Pacific region accounts for 15.3 trillion m³ (8.2%), Africa holds 14.2 trillion m³ (7.6%), North America has 12.1 trillion m³ (6.5%) and Central and South America together have 7.7 trillion m³ (4.1%). The United States is the world's largest user of natural gas and the rate at which it is consuming its reserves is faster than in any other region. Based on the BP figures in Table 2.2 current reserves would last for 12.8 years at the present rate of consumption.[6] In contrast African reserves can be expected to last for 69.6 years at the current rate of exploitation. In practice, the size of national and regional proved reserves is continually being updated so that these lifetimes are likely to be more extensive than the figures in Table 2.2 indicate.

Table 2.3[7] contains figures for the annual production of natural gas. This has risen steadily over the past 45 years. In 1970 annual production was 992 billion m³. This had increased to 1435 billion m³ by 1980 and by 1990 the global output was 1983 billion m³, double the production 20 years earlier. Total output reached 2416 billion m³ in 2000 and has continued to rise, year on year, with the exception of 2009 in the midst of the global recession when output was actually lower than in 2008. Output in 2010 more than made up for the earlier drop in output and in 2014 global annual production of natural gas was 3461 billion m³.

Table 2.4[8] expands on the figures from the previous table by showing the production and consumption of natural gas by global region in 2014. The largest regional production was in Europe and Eurasia where annual production was 1002.4 billion m³. The output was balanced by the largest regional consumption, 1009.6 billion m³. Much of this gas was burnt in Western Europe although Russia was also a large consumer. The production in Europe and Eurasia was almost

---

[6]According to the US Energy Information Administration, technically recoverable natural gas reserves in the United States were 64 trillion m³ at the end of 2012, sufficient if they could be recovered to last for 87 years. However this figure includes a large proportion of unproved resources.
[7]BP Statistical Review of World Energy June 2015.
[8]BP Statistical Review of World Energy June 2015.

Table 2.3 Global Annual Production of Natural Gas	
Year	Annual Production (Billion m^3)
1970	992
1980	1435
1990	1983
2000	2416
2001	2483
2002	2532
2003	2624
2004	2711
2005	2789
2006	2893
2007	2968
2008	3073
2009	2989
2010	3203
2011	3316
2012	3380
2013	3409
2014	3461
Source: BP.	

Table 2.4 Production and Consumption of Natural Gas by Region, 2014		
Region	Annual Production of Natural Gas in 2014 (Billion m^3)	Annual Consumption of Natural Gas in 2014 (Billion m^3)
North America	948.4	949.4
Central and South America	175.0	170.1
Europe and Eurasia	1002.4	1009.6
Middle East	601.0	465.2
Africa	202.6	120.1
Asia Pacific	531.2	678.6
Global total	3460.6	3393.0
Source: BP.		

matched by that in North America where output was 948.4 billion m^3. Consumption was virtually identical at 949.4 billion m^3. The Middle East was the third largest producer of natural gas in 2014 with 601.0 billion m^3. In contrast, consumption in the region was only 464.2 billion m^3. Large volumes of natural gas are exported from the

Middle East as liquefied natural gas (LNG). Much of this gas is exported to the Asia Pacific region. Here production in 2014 was 531.2 billion m³ whereas consumption was much higher at 678.6 billion m³. Africa produced 202.6 billion m³ of natural gas in 2014 but only consumed 120.1 billion m³. Large volumes of African gas are also exported. Finally, production of natural gas in Central and South America was 175.0 billion m³ while consumption was 170.1 billion m³.

Annual production from the top 10 natural gas producing countries in 2014 is shown in Table 2.5.[9] Top of the table is the United States with 728 billion m³, 21.4% of global output, followed by the Russian Federation with 579 billion m³, 16.7% of the global total. These two countries far outstrip all others. The next nearest rivals in terms of output are Qatar with 177 billion m³, Iran with 173 billion m³ and Canada with 162 billion m³. Other major producers in the table include China (135 billion m³), Norway (109 billion m³), Saudi Arabia (108 billion m³), Indonesia (73 billion m³) and Turkmenistan (69 billion m³).

The figures in Table 2.6[10] show the top 10 natural gas consuming nations. As with the previous table, the United States is at the top with 759.4 billion m³, followed by the Russian Federation with 409.2 billion m³.

Table 2.5 Top 10 Natural Gas Producers, 2014		
Country	Annual Production (Billion m³)	Proportion of World Total (%)
United States	728.3	21.4
Russian Federation	578.7	16.7
Qatar	177.2	5.1
Islamic Republic of Iran	172.6	5.0
Canada	162.0	4.7
People's Republic of China	134.5	3.9
Norway	108.8	3.1
Saudi Arabia	108.2	3.1
Indonesia	73.4	2.1
Turkmenistan	69.3	2.0
The rest of the world	1170	33.6
Source: BP.		

[9]BP Statistical Review of World Energy June 2015.
[10]BP Statistical Review of World Energy June 2015.

Table 2.6 Top 10 Natural Gas Consumers, 2014	
Country	Natural Gas Consumption (Billion m³)
United States	759.4
Russian Federation	409.2
China	185.5
Iran	170.2
Japan	112.5
Saudi Arabia	108.2
Canada	104.2
Mexico	85.8
Germany	70.9
United Arab Emirates	69.3
Source: BP.	

Third in the table is China with 185.5 billion m³, then Iran with 170.2 billion m³, Japan with 112.5 billion m³, Saudi Arabia with 108.2 billion m³, Canada with 104.2 billion m³, Mexico (85.8 billion m³), Germany (70.9 billion m³) and the United Arab Emirates (69.3 billion m³). Several of the countries in this table subsidize the cost of natural gas for their own populations and this pushes consumption much higher. These include the Russian Federation, Iran, Saudi Arabia, Mexico and the United Arab Emirates. In contrast the consumption in countries such as the United States, China, Japan and Germany is more closely matched to their economic capacities and outputs.

2.3 NATURAL GAS TRADE

The global trade in natural gas has two components, the trade in gas through natural gas pipeline networks and the trade in LNG. Much of the trade in pipeline gas takes place within countries such as the United States but there are also major international natural gas pipelines that transport gas across national borders.

National pipeline networks carry natural gas from producers to consumers. Within a single nation the management of such networks is relatively straightforward both technically and economically. However, the complexity of management rises when a pipeline crosses a border and involves two jurisdictions with different regulations. The situation can become even more complex when a pipeline carrying gas from one country to a second passes through a third, intervening country on the way. This latter country will make transit charges and may require some of the gas for its own use. Political bargaining can also play as part, as has happened in recent year with exports of gas from Russia to Europe through Ukraine.

In consequence of these difficulties most cross-border natural gas pipelines cross only a single border and have clearly defined exporters and final destinations for the gas. The major exception to this is the network of pipelines that connect Eastern and Western Europe. Most of these deliver Russian gas to Western Europe but future pipelines may also deliver gas from central Asian gas fields.

The main pipeline exporters of gas are the Russian Federation, Norway and Algeria. According to the IEA, the Russian Federation exported 203 billion m^3 of natural gas in 2013, Norway exported 103 billion m^3 and Algeria 45 billion m^3. All three sell gas to markets in Western Europe. Big importers of the gas include Germany which takes much of its gas from Russia and both Italy and Spain which import gas via pipeline from Algeria. Other European nations such as France and the United Kingdom are also big importers of natural gas.

The second natural gas export market is based on LNG. This usually involves natural gas being transported by pipeline to a port where it is liquefied and then loaded into a bulk tanker. The process is reversed at the destination port with the LNG being regasified before being delivered into a pipeline network for transport to consumers.

There have historically been two LNG trading regions, the Atlantic Basin and the Pacific Basin. The Atlantic basin in dominated by the key European markets of Italy, France, Belgium, the Netherlands and the United Kingdom. The main Pacific Basin importers, meanwhile are Japan, South Korea and Taiwan with the markets expanded more recently with the addition of China, India, Thailand and Singapore. Japan and South Korea between them accounted for 52% of global LNG imports in 2012. In recent years the separation of the two regions has become less clear with some exporters of LNG selling into both trading regions.

The number of LNG exporters has been growing steadily over the past 20 years as markets have expanded. In 1996 there were only

8 exporters[11] and 17 in 2012. The most significant is Qatar which exports LNG to both basins and in 2012 supplied close to one-third of global LNG supply. Other important LNG exporters include Malaysia, Australia, Nigeria, Indonesia, Trinidad, Algeria and Russia.

Much LNG is traded under bilateral deals but there is a growing spot and short-term market for LNG. In 2012 these accounted for 73.5 Mtonnes of LNG, or 31% of the total global LNG trade of 238 Mtonnes.[12]

[11]Figures are from www.natgas.info.
[12]IGU World LNG Report—2013 Edition, International Gas Union.

CHAPTER 3

Gas-Fired Power Generation Technology

There are a number of ways in which natural gas can be used to generate electricity. The first and simplest method is to burn the gas to generate heat and produce steam in a steam-generating boiler, then use the steam to drive a steam turbine. This type of power plant was originally developed at the end of the 19th century to burn coal and when natural gas became available during the first decades of the 20th century the steam-raising fossil fuel plant was adapted for gas use too. There is no definitive record of the earliest use of natural gas in this way but the first plants are likely to have been in the United States where the natural gas industry took off alongside the oil industry. Since then this type of power plant has continued to be used where natural gas is plentiful and particularly where there is no other immediate use for the fuel. Plants can be found in most natural gas producing countries, including the United States and Russia and in the Middle East, Africa and Asia. New plants of this type are still being built. However they are not the most efficient means of generating electricity from natural gas and these steam-raising plants using gas-fired boilers form only a small part of the global natural-gas-fired power generating capacity.

The inception of the most important way in which natural gas is used to generate electricity, in a gas turbine-based combined cycle power plant, is more clearly recorded. According to GE Power Systems, the first ever gas turbine installed in an electric utility in the United States, in 1949, was used in combined cycle mode.[1] This was not, however, the first use of a gas turbine for power generation. The credit for that belongs to the Swiss company BBC Brown Boveri which in 1939 installed a 4 MW open-cycle gas turbine at a municipal power station at Neuchâtel, Switzerland. This unit had a long life. It was finally retired in 2002 and has since been restored for public display.

---

[1]Combined Cycle Development, Evolution and Future, David L. Chase, GE Power Systems, 2000.

Gas-Turbine Power Generation. DOI: http://dx.doi.org/10.1016/B978-0-12-804005-8.00003-3

Open-cycle gas turbine plants, in which a free-standing gas turbine is used to generate electricity, were built during the middle of the 20th century in many developed countries where gas was available. They were relatively simple and reliable and in addition could be started up and shut down quickly which made them extremely useful both for meeting peak demand and for taking up demand during system failures. A very small-scale version of the open-cycle gas turbine, called a microturbine, has also been developed and can be used to supply power for domestic or small commercial use.

The first recorded combined cycle power plant to operate was at the Belle Isle Station, Oklahoma, United States where the first commercially sold gas turbine for power generation was installed in 1949 by GE.[2] The gas turbine had a generating capacity of 3.5 MW and the exhaust from the turbine was used to heat the feed-water for a conventional steam-generating unit, probably fired with coal. This unusual design feature was a result of restrictions and procurement difficulties in the United States after the Second World War. The plant had more turbine capacity than boiler capacity and the additional heat input allowed more of the steam turbine's capacity to be harnessed. The heat recovery heat exchangers (called economizers) used in the plant were bare tubes, placed in the exhaust path, through which the feed-water passed. This design was typical of early combined cycle heat exchangers.

Further early combined cycle plants entered service during the 1950s and early 1960s. These often used the exhaust from the gas turbine as the combustion air for a conventional fossil fuel-fired boiler. It was only when more advanced heat recovery boiler tube designs were developed that designs more reminiscent of a conventional combined cycle plant began to appear. Finally, during the early 1980s, the modern combined cycle power plant started to attract the attention of power companies and the technology began to be adopted widely.

Beside gas turbines, another way of utilizing natural gas to generate electricity is in a piston engine. These engines also have a long history of power generation and have been used to supply mechanical power for gas compression and pumping stations in the natural gas industry

---

[2]The name combined cycle power plant is a relatively recent terminology. Early plants would not have been called combined cycle stations.

for decades. Used to drive a generator, natural gas engines of this type are efficient and clean and have become popular for small-scale distributed generation applications.

There is one further important means of using natural gas to provide electricity, in a fuel cell. Fuel cells generally require hydrogen to operate. It is possible to convert natural gas into hydrogen and use the product to provide fuel for a fuel cell, with the natural gas converter (usually called a reformer) and the fuel cell integrated into a single unit. As with piston engines, fuel cells are generally used for small-scale distributed generation rather than as central power plants.

3.1 NATURAL GAS-FIRED STEAM TURBINE POWER PLANTS

The workhorse of the power generation industry for much of the 20th century and at the start of the 21st century is the coal-fired power station. This type of power station is designed to burn coal in a controlled fashion and use the heat generated to raise steam and drive a steam turbine. The same type of power plant can be utilized to burn either natural gas or oil.[3] However most of the developments in steam plant boiler and turbine technology have been driven by their use in coal-fired power plants. So while plants that burn natural gas or oil can be found where there were plentiful supplies of these fuels, the design of these plants has generally been based on that of a similar coal-fired plant. This synergy continues today.

In consequence, a natural gas-fired boiler for power generation is very similar to a coal-fired boiler and in many cases the two types of boiler from one company will share components. The layout of a typical gas-fired steam plant is shown in Fig. 3.1. In the case of a gas-fired plant, natural gas[4] is burned in a controlled amount of air in a furnace. The heat generated from the combustion is then captured by water flowing through pipes within the walls of the furnace and in specially designed pipe bundles that are placed in the path of hot gases exiting

[3]Liquid or gaseous fuel is much easier to handle than coal and plants that burn oil can usually burn natural gas, and vice versa. Many natural gas plants of all types are designed for dual-fuel operation.
[4]Or oil. Many of these plants are designed to burn both natural gas and oil. Although combustion conditions and handling techniques will be slightly different, both are much easier to manage than coal and many of the natural gas-fired steam-raising plants being built are capable of burning oil too.

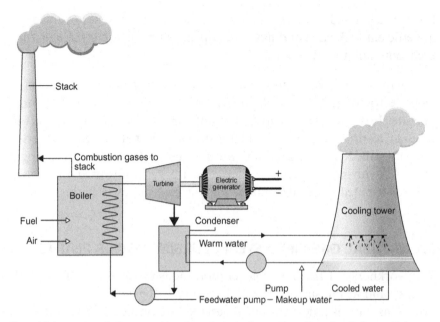

Figure 3.1 A natural gas-fired steam turbine plant schematic. Source: From Zhengzhou Boiler co., Ltd.

the furnace. The ideal is to capture as much of the heat generated as possible to raise steam so that the air exiting the power plant boiler chamber should be at as low a temperature as possible.

Early steam boilers of this type produced relatively low-temperature, low-pressure steam but as materials technology has developed it has become possible to build plants capable of generating much higher pressure, higher temperature steam. Temperatures in coal-fired boiler furnaces can now reach 1700°C. The temperature in a gas-fired furnace will probably be somewhat lower than this.

The driving force behind the quest for every more extreme conditions of temperature and pressure is efficiency. The energy conversion process by which thermal energy in the natural gas is converted into electrical energy involves a steam turbine. This is a type of thermodynamic heat engine and its overall efficiency is limited by the Carnot Cycle established by the French scientist Sadi Carnot during the 1820s. What the Carnot Cycle teaches is that the maximum efficiency of which a heat engine is capable is determined by the temperature drop that the working fluid—in this case steam—experiences between entering the engine, the steam turbine and exiting it.[5] Since the temperature at the exit of the steam turbine can usually be no lower than ambient temperature, which in most cases is fixed, the only means of increasing overall efficiency is to raise the temperature of the steam that enters the steam turbine.

Water in the form of steam is the working fluid for a steam turbine and its physical properties play a crucial role in boiler design. Water at normal temperatures and pressures behaves as we all expect it to. It is a liquid at room temperature but as the temperature is raised it eventually boils. The boiling point is 100°C at normal pressures but will rise as the pressure of the water rises. However if the pressure is raised far enough—above 22.1 MPa (221 times atmospheric pressure)—the nature of the water changes and it becomes what is known as a supercritical fluid. Once it enters this phase there is no difference between the liquid and the gaseous version of the fluid and the two coexist. The supercritical point of water is actually defined as 22.1 MPa/374.1°C since it is dependent on temperature too; above this point water is a supercritical fluid and below it a normal fluid.

Most power plant boilers designed before the latter part of the 20th century, be they for natural gas, oil, or coal combustion, operated at below the critical point because the materials needed to build reliable boilers for higher temperatures and pressures were not available. These types of boiler are referred to as subcritical boilers. Many modern power plant boilers now operate with steam above the critical point and these are called supercritical boilers. A further class, called Ultra-supercritical, operate under even more extreme conditions but are usually only used for coal combustion.

Modern natural gas-fired boilers for power plants can have capacities of up to 1000 MW, although most are typically smaller than that, and they can operate with steam pressures of up to 25 MPa and steam temperatures of over 550°C. The furnaces of these plants are designed to control the combustion conditions in order to limit the production of nitrogen oxide. This is achieved by controlling the amount of air allowed to enter the furnace during the early, highest temperature stages of the combustion process when most nitrogen oxides are produced.

[5]When calculating Carnot efficiencies, all temperatures are measured in degrees Kelvin relative to absolute zero.

Natural gas contains only traces of nitrogen but the nitrogen in air that is introduced into the furnace during combustion will be oxidized at the very high temperatures in the furnace fireball. Limiting the amount of oxygen available for combustion at this stage limits the production of oxides of nitrogen because the oxygen that is available preferentially reacts with methane. Further air is then added higher up the combustion chamber where the combustion gases have cooled slightly to allow the combustion to continue to completion at lower temperature and with less danger of nitrogen oxidation. For natural gas-fired supercritical boilers of this type the efficiency can be as high as 49%. It will be lower for subcritical boilers and smaller units are also likely to have lower efficiency than the largest units.

Natural gas is a relatively clean fuel compared to coal and only limited flue gas cleanup will be required before the flue gases can be released into the atmosphere. The only major pollutants likely to be present at a significant concentration are the nitrogen oxides discussed above, usually referred to collectively as NO_x. Where the level of NO_x exceeds local environmental limits a removal system will have to be used. Various technologies are available that will convert the nitrogen oxides back to nitrogen. The most common of these are selective catalytic reduction and selective noncatalytic reduction. Both use a reagent such as ammonia or urea to react with the NO_x. Similar technologies are used for other types of natural gas-fired power plant (see Chapter 9).

The other main environmental concern is with carbon dioxide. A natural gas-fired boiler will generate significant quantities of carbon dioxide during the combustion process. While the amount per unit of electricity will be smaller than for a similar coal-fired power plant, future legislation is likely to require such plants to control their carbon dioxide emissions by capturing and storing the gas.

The steam from the power plant boiler is used to drive a steam turbine that is matched to the output of the boiler. As with the boiler, this is likely to be a turbine designed primarily for a coal-fired plant operation. Depending on the plant size it may comprise high pressure, intermediate pressure and low-pressure turbine units together will the possibility of reheating steam between one or more of these units to improve overall efficiency. The steam turbines will be coupled to generators that convert the rotary motion from the turbine shaft into electrical energy.

Gas-fired steam plants of this type are only likely to be found in regions where there is oil production but no ready market for the associated natural gas. Natural gas is a valuable product and it will normally be sold unless local conditions make this difficult. Typical locations for such plants are the Middle East, in African oil producing countries, and in Asia. Natural gas is unlikely to be used in this way in Europe or the United States today.

3.2 PISTON ENGINE-BASED NATURAL GAS POWER UNITS

The piston engine or reciprocating engine has a long history in the power generation. Some of the very first coal-fired power stations that were built in the 19th century used steam reciprocating engines to drive generators. Modern reciprocating engines are used mainly for transportation. Small engines are used in domestic vehicles and larger ones in trucks, locomotives and ships. Equivalent engines can be adapted for the power generation market. In terms of power output, sizes can range from as small as 0.5 kW to as large as 65 MW.

There are two main categories of piston engine suitable for power generation, spark ignition engines and compression ignition engines, but only the first of these can be fired with natural gas. Compression ignition engines are usually fired with diesel. There are also different cycles under which a piston engine can operate. The two most common are the two stroke and the four stroke engine. Engines using both types of cycle can be operated with natural gas.

A further variable is the ratio of air to fuel within the combustion chamber (the cylinder) of an engine. Some operate with a roughly stoichiometric ratio of oxygen from air and fuel, such that there is just sufficient oxygen for all the fuel to burn. Such engines are referred to a rich burn engines. These engines tend to operate at high combustion temperatures and this can lead to the production of relatively high levels of nitrogen oxides as well as other pollutants. The alternative is a lean burn engine in which there is far more air (and oxygen) than is required for combustion. The excess air leads to lower combustion temperatures within the engine cylinders and lower pollutant levels in the engine exhaust. Under normal circumstances a rich burn engine will generally provide higher efficiency than a lean burn engine. However modern design of lean burn engines is allowing them to reach similarly high levels of efficiency while maintaining lower emissions production levels.

As with natural gas-fired steam turbine plants, the main environmental consideration is NO_x. Rich burn engines burning natural gas will normally require some form of catalytic reduction system to remove NO_x and bring the emissions level within local regulations. Some lean burn engines may be able to meet environmental regulations without the need for additional emission control systems. Engines also generate carbon dioxide but it is unlikely that it will be cost effective to apply carbon capture technology to piston engines except in the very largest of installations.

Natural gas-fired piston engines are available in sizes from 0.5 kW to around 6 MW. For power plants larger than this multiple engines are usually required. While larger piston engines can be built, these normally operate on heavy oil as fuel rather than natural gas. The speed of a piston engine varies depending on its size. Natural gas engines can be either high speed engines (1000–3000 rpm), which are available in sizes from 0.5 kW to 6 MW or medium speed engines (275–1000 rpm) which usually start at 1 MW. The larger, slower speed engines tend to be more reliable and are typically chosen for continuous operation. Where intermittent operation is required, smaller, high speed engines will often be chosen because they tend to be cheaper, though less reliable.

The uses for natural gas-fired engines for power generation are varied. Many of them are used for distributed generation applications where they supply power directly to local consumers. Some of these engines are used in cogeneration mode in which waste heat from the engine is used to heat water. This can lead to very high overall efficiencies. Another common application is for grid backup, with systems designed so that they start up as soon as there is an interruption to the mains supply. Natural gas engines may also be used in conjunction with renewable capacity, such as wind power or solar power, in microgrid type applications where they are also used as a standby power supply.

3.3 FUEL CELLS

A fuel cell is a form of electrochemical cell, similar to a battery. Batteries exploit the energy that is released in a spontaneous chemical reaction, turning the energy that would normally be released in the form of heat into electrical energy. The exact chemical reaction varies from battery to battery and in the case of the fuel cell it is the reaction between hydrogen and oxygen to create water.

In order for a fuel cell to operate, hydrogen must be fed to one of its electrodes and oxygen to the other. The electrodes are coated with catalysts that encourages molecular hydrogen or oxygen—depending upon the electrode—to split into its atomic components, two hydrogen atoms or two oxygen atoms. These atoms are extremely reactive but are at different electrodes of the cell. The cell is designed in such a way that in order for them to react with one another, a hydrogen atom must give up an electron, forming a positively charged hydrogen ion and this electron must then be passed to an oxygen atom to create a negatively charged oxygen ion.[6] The electrolyte between the electrodes will then allow either the charged hydrogen or a charged oxygen ion (but not both) to pass from one electrode to the other so that the two can react. However this electrolyte will not conduct electrons so these can only be transferred via an external electrical circuit. In order for the chemical reaction to be completed, charged atoms move through the electrolyte while electrons that keep the reaction balanced travel through the external circuit. In this way the fuel cell turns the chemical energy that drives the reaction into electrical energy.

There are a number of different types of fuel cell available today, each defined by the type of electrolyte it utilizes. The most common are the phosphoric acid fuel cell, the proton exchange membrane fuel cell, the molten carbonate fuel cell and the solid oxide fuel cell. Methanol fuel cells are also being developed for small-scale applications. Fuel cells are clean, environmentally friendly generation systems that can be installed in urban areas without problem. When running on natural gas a fuel cell power plant will produce carbon dioxide but they do not generate significant levels of any other atmospheric pollutants. However most fuel cells for power generation applications are relatively expensive.

[6]The oxygen ion may in fact be the hydroxyl OH^- ion.

CHAPTER 4

Gas Turbines

While the technologies discussed in Chapter 3 all offer important ways of using natural gas to produce electricity, it is the gas turbine that has revolutionized the use of natural gas for the power industry. The modern gas turbine was originally developed as an aero engine but was quickly adapted to stationary applications including power generation. However, initial uptake was limited. It took several decades before modern, heavy duty gas turbines evolved and became the dominant type of gas turbine in the power generation market.

Like most power generation technologies, the technological principles upon which the gas turbine is based were first explored during the 19th century. Working machines were first perfected during the early years of the 20th century but it was the recognition of their potential as aviation power units that accelerated the development of the technology. Power generation applications were tested as early as 1939 but the use of gas turbines in the power generation industry only picked up gradually after the Second World War and until the 1980s most gas turbines for power generation were relatively small.

The potential for high efficiency combined cycle plants was fully recognized during the 1980s and this encouraged manufacturers to develop large, heavy duty gas turbines specifically for this market. This led, in turn, to rises in the overall efficiency of such power stations so that in the second decade of the 21st century power plants with overall efficiencies of over 60% are in operation. Gas turbines in simple open cycle configuration can reach 46% efficiency.

4.1 THE HISTORY OF THE GAS TURBINE

A gas turbine is a device that converts the energy contained in a gas, usually air, into rotary motion. One device of this type is the windmill

Gas-Turbine Power Generation. DOI: http://dx.doi.org/10.1016/B978-0-12-804005-8.00004-5

or wind turbine which converts the energy of moving air, the wind, into rotary motion. While the wind turbine harnesses the energy of free air, a gas turbine is normally driven by a high pressure, often high temperature gas that is constrained within a casing. This allows much greater control of the energy capture process.

The very earliest forerunner of this type of device was built by Hero of Alexander in the 1st century AD.[1] It used the reaction of jets of steam from a pair of nozzles on opposite sides of a boiler vessel mounted on a shaft to drive rotation of the vessel about its axis. This use of hot, expanding gas to drive motion is common to both steam and gas turbines. Another forerunner of the gas turbine is the "chimney jack." This was a simple turbine that was mounted into a chimney so that hot air rising up the chimney caused it to turn. This motion could then be harnessed, for example, to turn meat on a rotating spit.

The first invention containing all the elements of a modern gas turbine was described by the English inventor John Barber in a patent taken out in 1791. This "Method of Rising Inflammable Air for the Purposes of Procuring Motion" contained a compressor, a combustion chamber and a turbine, the key elements of any modern gas turbine. It is not known if the machine was ever built or operated successfully but modern models have been built and do produce motion.

During the 19th century a number of early gas turbines were developed. These all used a compressor to generate a flow of pressurized air and this flow was fed into a turbine in order to drive a shaft and produce mechanical work. In all these machines the compressor was separate from the turbine and while overall efficiency was probably very low, the fact that the compressor was decoupled from the turbine meant that they were always able to produce power to drive machinery.

The first attempt to design what is recognizably a modern gas turbine can be found in a design by the German engineer Franz Stolze, published in 1872. The design used an axial compressor to compress air that was then fed into a combustion chamber with a fuel which was ignited to produce a stream of very hot, high pressure gas. This hot gas was fed into a multistage turbine which rotated, driving a shaft.

---

[1]An aeolipile was described by Vitruvius more than a century earlier but from the description it is not clear if he is talking of the same machine.

Crucially, both the compressor and the turbine were mounted on the same shaft, as in a modern turbine. In order for a design of this type to be able to run, the turbine must produce more power than is needed to drive the compressor. Otherwise it cannot operate continuously. The Stolze design was never able to do this because neither turbine nor compressor was efficient enough for continuous operation to be possible. That requires a minimum of around 80% efficiency for each part of the system.

The first gas turbine that was capable of sustained operation was built by Norwegian engineer Aegidius Elling and demonstrated in 1903. The Elling turbine produced power in excess of that needed to drive its components and had a net output of 8 kW which was delivered as compressed air that could drive pneumatic machinery. The design also incorporated water injection into the hot gases between the combustion chamber and the turbine, a process that is used today in some modern gas turbine cycles. The gas turbine inlet temperature was 400°C, low by modern standards. Turbine speed was 20,000 rpm. In 1904, Elling was able to increase the inlet temperature to 500°C and the power output to 33 kW.

There were further developments during the first part of the 20th century. However the next milestones occurred during the 1930s when Swiss company BBC Brown Boveri started to manufacture axial compressors and turbines for supercharged steam generation. At the same time the British engineer Frank Whittle was developing a gas turbine for jet propulsion. His first engine ran in 1937 while in 1939 Brown Boveri installed a gas turbine in a power station for the first time as a generating unit. This marked the birth of the new power generating technology.

4.2 THE GAS TURBINE PRINCIPLE

A modern gas turbine has three key components, a compressor, a combustion chamber and a turbine as shown in Fig. 4.1. Air is drawn into the compressor and compressed, then fed into the combustion chamber where fuel is injected continuously, releasing heat which raises both the temperature and the pressure of the air. This high temperature gas stream is then fed into the turbine to produce power. In the standard design the compressor and combustion chamber are mounted on the

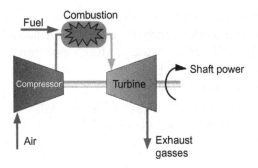

Figure 4.1 The gas turbine principle. Source: From Wikimedia.

same shaft and that shaft is also coupled to the generator. The turbine stage of the machine must therefore generate enough shaft power to both turn the compressor and rotate the generator.

The turbine section of a gas turbine is a type of heat engine and so like the steam turbine discussed above in Chapter 3, its energy conversion efficiency is determined by the Carnot Cycle efficiency-limit which depends on the difference between inlet and outlet temperatures. This means that the inlet gas temperature is a key parameter for determining overall efficiency and much of the development of modern gas turbines has been focused on finding materials and techniques that allow ever higher inlet temperatures to be achieved.

The actual cycle upon which a gas turbine operates is called the Brayton Cycle after George Brayton who first conceived it. Its defining feature is that heat is added to the working fluid, air, at constant pressure. At the time of its invention, the Brayton Cycle could not compete with the reciprocating Otto Cycle engine but the development of the gas turbine engine, which turned out to operate on the Brayton Cycle—though this was not immediately recognized—showed that it could be extremely effective.

Brayton's original engine, which was a development of the work of John Barber, was a type of piston engine with one piston to compress air which then passed into a mixing chamber where fuel was added. This mixture was then ignited as it entered the expansion chamber where it drove another piston. The compression cylinder piston was driven by the expansion chamber piston.[2] The gas turbine cycle shares similar compression, combustion and expansion sections, although they operate completely differently.

The early development of gas turbines was aimed at producing jet engines for military use. This work, which built on that of Whittle, carried on through the 1940s, 1950s and 1960s. Aero engine development continues, though today it is as often for commercial as for military aviation applications. A cross-section of a gas turbine jet engine is shown in Fig. 4.2.

A jet engine takes air in through its front air intake, compresses it, then burns fuel in the compressed air to create a hot high pressure jet. This high pressure air passes through a set of turbine blades where some power is extracted to drive the compressor. The remaining blades reduce the gas pressure and increase the velocity of the air which is expelled from the rear of the turbine. This jet of air creates the thrust that, by Newton's third law of action and reaction, is used to get the aircraft airborne and keep it there. The turbine may be conceived as a propeller enclosed in a shell instead of a propeller that turns in free air as was more traditional with early aeroplanes. Enclosing the propeller allows the input air speed to be controlled independently of the rate at which the machine moves through the air and this allows the aeroplane to travel faster.

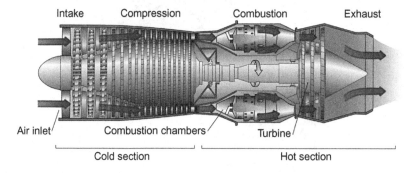

Figure 4.2 Schematic of a gas turbine aero engine. Source: From Wikimedia.

[2]In a normal piston engine, such as an Otto Cycle engine, the compression, ignition and subsequent expansion stages all take place within a single cylinder.

The adaptation of these turbines for mechanical drive and power generation meant redesigning the turbine stages of the gas turbine so that as much energy as possible is captured from the hot, high pressure air that is delivered into the turbine from the combustion chamber. Instead of creating thrust, the energy is delivered to the turbine shaft as mechanical power. These aeroderivative gas turbines proved to be efficient and highly cost effective and they still form a major part of the gas turbine industry today.

Aero engines must be light in order to reduce overall airborne weight. Weight is less of a consideration for stationery and industrial use. This has allowed aeroderivative gas turbines to diverge from their aero engine heritage although many retain close similarities. In the last decades of the 20th century the major gas turbine manufacturers began to adapt the gas turbine even further, building heavy duty gas turbines that were specifically for power generation. The cross-section of a large industrial gas turbine is shown in Fig. 4.3. These units have since become much bigger than their aeroderivative counterparts and design philosophies have also diverged. Even so, aero engine developments, particularly related to design of new materials, quickly find their way into the larger industrial machines.

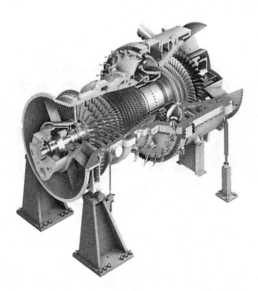

Figure 4.3 Cross-section of a large industrial gas turbine. Source: With permission from Siemens. Siemens SGT5-4000F 300MW gas turbine.

4.3 A NOTE ABOUT PRESSURE

The operation of the components in a gas turbine depends on the interaction of these components with high pressure air, an interaction which involves changes in the pressure of the air. There are two different types of pressure that need to be considered when analyzing how these components operate, static pressure and impact, or dynamic pressure. Static pressure is the pressure that a stationary fluid exerts on its surroundings. For example, the air in a balloon exerts a static pressure on the fabric of the balloon. Dynamic pressure on the other hand is the pressure that is created by a moving fluid. The pressure or force exerted by a jet of water is an example of dynamic pressure. When a fluid such as air is moving it will exert both a dynamic and a static pressure. These two together make up the total pressure of the fluid. In the absence of any energy input or loss, the total pressure remains constant. However the balance between static and dynamic pressure can change. This is important for gas turbine operation.

The interaction of the components in a gas turbine with the working fluid, air, affects the balance between the two sorts of pressures. The shape of the air paths within the gas turbine or compressor can convert dynamic pressure into static pressure and vice versa. If a stream of high speed air passes through a duct that diverges so that its cross-sectional area increases, the high speed air will fill the increasing volume. This will decrease its velocity but since there is no energy change taking place, the static pressure must increase in order for the total pressure to remain unchanged.[3] A converging duct will have the opposite effect, increasing the velocity of the fluid flowing through it. Both these effects take place in gas turbine compressors and turbines.

The processes that take place in a gas turbine also involve changes in temperature. Compression of a gas leads to an increase in its temperature while expansion leads to a fall in temperature. Heat energy is added both in the compressor, during compression, and in the combustor. The heat energy is then expanded during the energy conversion process within the turbine.

---

[3]This ignores frictional losses.

4.4 COMPRESSORS

The role of the compressor in a gas turbine is to produce a stream of highly compressed air that can be fed into a combustion chamber where it is mixed with fuel and then ignited. During the compression process the pressure of the air increases, but its velocity as it passes through the machine should not. The compressor is normally mounted onto the same shaft as the turbine and uses some of the power developed by the turbine to carry out its function.

The compressor of a modern gas turbine will typically have 10 to 16 sets of compressor blades and vanes, usually called stages. Each stage will compress the air from the previous stage. The overall compression ratio varies from manufacturer to manufacturer and from turbine to turbine but will typically be between 15:1 and 30:1.

Most gas turbines use axial flow compressors. In such a compressor each stage consists of a set of blades that are attached to the shaft of the compressor and rotate with it, followed by a set of static vanes. The blades in the first stage draw in air and increase its pressure and velocity. This air then passes through the static vanes which by their shape convert the increase in velocity into an increase in pressure. The air then passes to the next stage where further compression takes place, and so on until the air exits the final stage at the required pressure. Compression also increases the air temperature so that the air exiting the compressor is hotter than the air that entered.

The compressor section of the gas turbine also includes inlet guide vanes to ensure that air is drawn into the first stage at the best angle. Finally there are also outlet guide vanes that direct the air into the combustion chamber.

Compressor efficiency is typically around 87%. Overall efficiency is determined by the smoothness of the air flow through the compressor and much of the design effort is to ensure that this is as smooth as possible. Blades and vanes are generally in the shape of aerofoils to maintain smooth flow and prevent turbulence. It is also important to prevent any backwards flow of air as the pressure increases from stage to stage along the compressor. In order to achieve this, the clearance between the tips of the rotating blades and the casing is kept to a minimum and the static vanes have seals to prevent backward leakage.

The precise shape and angle of blades and static vanes depends on the rotational speed of the compressor so the machine needs to operate at its design speed in order to achieve the highest efficiency.

4.5 COMBUSTION CHAMBERS

The combustion chamber of a gas turbine is where the energy that drives the whole system is added. The combustion chamber of a modern turbine typically consists of a cylinder with a second smaller cylinder called the liner inside it. A fuel air mixture passes into the mouth of the liner and additional air may pass around the outside of it, between the liner and the outer cylinder to keep the liner cool. This air is then introduced through holes and slots along the liner.

In most modern gas turbine combustors the air is premixed with fuel before it is injected into the combustion chamber through a set of nozzles. The shape and direction of the nozzles and baffles in the combustor are carefully designed to ensure both even mixing and a stable flame within the combustor. The fuel air mixture ignites in the combustion zone, releasing energy as heat. The temperature in the combustion zone flame can reach over 1900°C, far higher than most materials can withstand. In order to control this some of the air from the compressor may be used to cool the walls of the combustion chamber liner. This will also dilute the very hot combustion gases to reduce their temperature.

The air flow through all parts of the combustion chamber must be carefully managed to avoid flame instability and turbulence which will lead to energy loss. The aim is to produce a smooth flow of air, even though the addition of heat energy will raise its temperature and increase the total pressure.

The addition of air into the combustion chamber is also carefully managed in order to control the production of NO_x during the combustion process. The high temperatures within the combustion zone will lead to ready production of nitrogen oxides from the reaction between oxygen and nitrogen from air. This can be controlled by maintaining reducing conditions. By keeping the amount of oxygen low compared to the quantity required to burn all the fuel, NO_x production can be minimized. With this type of staged combustion further air is introduced into the latter stages of the combustion zone in order to allow the combustion reaction to continue to completion. However many modern combustors rely on careful mixing of the fuel and air in stoichiometric proportions before the mixture enters the combustor to keep NO_x production under control.

After the combustion process is complete the hot gases pass into the final stage of the combustion chamber which is called the transition section. This is a convergent duct that will convert static pressure into dynamic pressure, increasing the velocity of the hot gases before delivering them into the turbine section.

The type and number of combustion chambers in a gas turbine will vary from manufacturer to manufacturer and from turbine to turbine. Many larger turbine designs use a set of annular combustion chambers that surround the turbine shaft between compressor and turbine. Others take air from the compressor outside the body of the turbine to one or more combustion chambers and then return the gases to the turbine.

At least one manufacturer of heavy industrial gas turbines also uses multiple sets of gas turbines and combustion chambers. This design splits the turbine section of the gas turbine into two. Hot air from the first set of combustion chambers enters the first turbine section where energy is extracted by the turbine blades, then the air enters a second set of combustion chambers where more fuel is burnt and more energy added before being fed into a second turbine section. This type of design, called a reheat turbine, is often used in large steam turbines for power generation but is much less common in gas turbines.

4.6 TURBINES

The final part of the gas turbine is the turbine section. This is where the energy from the fuel is converted into a form of mechanical energy, the rotation of the turbine shaft generating a torque. All gas turbines except for some very small machines use axial flow turbine sections. As with the compressor, the axial flow turbine will consist of a number of stages, each stage including a set of stationary vanes, usually called nozzles and a set of rotating blades that are attached to the turbine shaft.

There are two major types of turbine/blade design that can be applied to a gas turbine, each defined by the way it extracts energy from a fluid. The two are called reaction turbines and impulse turbines. One way of understanding the difference is to observe that reaction turbines exploit the static pressure in a fluid whereas impulse turbines exploit the dynamic pressure. This means that as fluid passes through a reaction turbine the static pressure drops but the fluid velocity which defines its dynamic pressure remains relatively constant. In contrast when a fluid passes through an impulse turbine stage the velocity drops while the static pressure remains constant. The stages of a modern axial gas turbine will generally combine the two, extracting part of their energy from the static pressure and partly from the dynamic pressure. It is common for the first stages to be predominantly impulse type while the latter stages are more reaction type. However all stages usually exploit both.

The order of the stationary vanes and rotating blades in the turbine is the reverse of the compressor. The high pressure, high temperatures gas from the combustor meets a stage's vanes first and then is directed to its blades. The vanes form convergent ducts which turn static pressure into dynamic pressure, increasing the speed of the air passing through them. This dynamic pressure is then used to drive round the rotating blades. As in the compressor, both vanes and blades are shaped like aerofoils in order to ensure the smooth flow of air through the complete turbine. Each stage extracts a portion of the energy contained in the air.

In a simple gas turbine the compressor and the turbine blades are all on a single shaft. However there are more complex arrangements. In some machines there are two concentric shafts. One of these carries the compressor blades and the first one or two stages of turbine blades. The later turbine stages are attached to a second shaft that drive the generator to produce electrical power. In some aeroderivative gas turbines this is taken further still and the compressor stages are divided too. The low pressure compressor blades are then mounted onto the same shaft as low (or medium) pressure turbine stages while the high pressure compressor stages are on the same shaft as the high pressure turbine stages.

The efficiency of a gas turbine will depend on the temperature drop across the stages. In order to achieve high efficiency the turbine stage inlet temperature must be very high. In some modern gas turbines the inlet temperature can reach 1600°C. It requires very special materials and design techniques to design turbine components that can withstand this temperature.

The efficiency of a gas turbine will depend not only on the inlet gas temperature but also on the temperature of the gas as it leaves the final stage of the gas turbine. The exhaust gas from a simple cycle gas turbine, that is one not in a combined cycle configuration, needs to be a cool as possible in order to achieve maximum efficiency. However in a combined cycle power plant a part of the energy is captured in a steam generator that exploits waste heat in the gas turbine exhaust. The temperature of the exhaust gas exiting a turbine in this type of plant will be much higher. The temperature at the exhaust of a high efficiency aeroderivative gas turbine will probably be in the 400°C to 500°C range. While this is relatively high it still enables efficiency of up to 46% for the best machines. Other small industrial turbines will have efficiencies of up to 42%. Conversely large industrial gas turbines designed for combined cycle operation may have exhaust gas temperatures above 600°C. Efficiencies may be as low as 38% but will typically be up to 42%.

CHAPTER 5

Advanced Gas Turbine Design

The evolution of gas turbines since the middle of the 20th century has involved the development of a range of advanced materials and the optimization of the three major components, compressor, combustor and power turbine. Early development was all centered around aero engines and its primary aim was to produce high power levels from gas turbines to enable ever faster flight. This required a combination of high efficiency and low weight. Materials and component design soon became key to achieving these aims. Both of these features carried over into early aeroderivative gas turbines that exploited the key properties of flexible operation and easy deployment to apply the technology to both mechanical drive and to the power generation industry.

Towards the end of the 1970s it became clear that the two arms of the gas turbine industry were diverging. While weight was an issue for jet engines in aircraft it was much less important for the power generation sector. In consequence stationary gas turbines did not need to use titanium-based alloys that were important for aero engines. Instead efficiency became the most important aspect. High efficiency for power generation depends on optimizing the design of all the components of the gas turbine; compressor, combustion chamber and power turbine to deliver shaft power.

By the 1990s the advent of computer-aided design coupled with mathematical modeling had allowed both compressor and power turbine component design to be reach a high level of optimization. While there will always be room for improvement, it was now much more difficult to increase efficiency by modifying the design of these components.

The only remaining means of increasing efficiency was by changing the operating parameters of the machine. The two main parameters available to the designer are the compression ratio of the engine, determined by the compressor, and the power turbine inlet temperature. The former is a design choice which varies from manufacturer to

Gas-Turbine Power Generation. DOI: http://dx.doi.org/10.1016/B978-0-12-804005-8.00005-7

manufacturer but remains relatively fixed within each design family. Typically the compression ratio is between 15:1 and 30:1 for modern gas turbines. That leaves power turbine inlet temperature. Raising this allows efficiency to be increased by increasing the overall temperature drop between inlet and outlet. It is in this area that most effort has been directed over the last 20 to 30 years.

Higher efficiency, obtained with a higher turbine inlet temperature, remains the key to improved performance of stand-alone gas turbines. However, for combined cycle power plants where power generation is shared between a gas turbine cycle and a steam turbine cycle, this must be combined with a balanced design. Extracting the maximum amount of energy and achieving the highest efficiency, which for the best systems is now in excess of 60%, demands careful integration of the two cycles and precise control of operating conditions. Moreover the best results are achieved when the system operates at a steady state and many of the best performing high efficiency combined cycle power plants operate under base load or intermediate load conditions.

In recent years, however, the duty cycle required of combined cycle plants has changed markedly, particularly in developed countries where the introduction of large volumes of renewable power into grid systems has become common. This has led to the need for combined cycle plants to be able to provide system back-up services involving much more frequent start-up and shutdown and the ability to ramp power both up and down rapidly. Achieving high efficiency under these conditions is much more difficult and major gas turbine companies now offer gas turbines and combined cycle plants that are capable of highly flexible operation. This comes at a price. Operating under frequency changing conditions will generally lead to a reduction in efficiency.

5.1 GAS TURBINE MATERIALS

Gas turbines operate at high temperatures and at extremely high speeds. This places great demands on all the components of the machine. In consequence designers and materials engineers have developed a complete range of specialist materials from which to build these components. Materials that are capable of providing great strength and durability at high temperatures are key among them.

As noted above, increasing the power turbine inlet temperature of a gas turbine represents the main means used by designers to achieve an increase in efficiency. An 8°C increase in the inlet temperature can lead to an increase in power output of 1.5% to 2.0% and an increase in efficiency of 0.3% to 0.6%.[1]

Inlet temperatures have risen steadily over the past century as materials have improved. The inlet temperature of the first gas turbine to operate successfully, built by Aegidius Elling in 1903, was 400°C, increasing to 500°C the following year. By 1967 the first stage inlet temperature of the most advanced gas turbines had reached 900°C and in 2000 temperatures of 1425°C were possible. Gains have continued into the 21st century and in 2011 Japanese company Mitsubishi Hitachi Power Systems developed an advanced gas turbine with inlet temperatures of 1600°C. Efforts are underway in Japan to push this to 1700°C in the near future. Beyond that, new targets will no doubt be set.

While the increase from 400°C to 1600°C represents an enormous advance in the space of almost 110 years, gas turbine combustors could deliver gas at a higher temperature still. The limiting factor remains the material from which the first stage of the power turbine is made. There are no modern materials that can withstand even the existing temperatures inside a combustor with ease and great technical sophistication has been required to be able to exploit the current highest inlet temperature, using a mixture of advanced cooling techniques and sophisticated materials technology.

Moreover, it is not just in the power turbine that high performance materials are required. The combustor liner must withstand extremely high temperatures too while the compressor stages, though not exposed to such high temperatures, have to be able to withstand corrosion from air laden with moisture containing salts and acidic materials that are drawn into the machine.

5.2 COMPRESSOR

The primary metal used for compressor blades in aero engines is titanium, in the form of a range of titanium alloys. The metal is preferred for its low weight and relatively high temperature resistance. Since the

[1]Power Generation Handbook, Chapter 14 Gas Turbine Materials, McGraw-Hill, 2004.

1950s the amount of titanium in an aero engine has increased from around 3% to 33%.[2] The best titanium alloys are resistant to up to around 540°C. However, this is too low for the last stages of the compressor of modern aircraft engines and these must be made from higher temperature nickel alloys which weigh almost twice the equivalent titanium component.

For stationary applications weight is not usually a consideration and the compressor blades and nozzles for heavy duty gas turbines and aeroderivative gas turbines can be made from steels. These steels usually contain chromium and carbon and some more recent ones contain nickel and molybdenum too. These steels have high tensile strength and high cycle fatigue strength. They are resistant to acidic salts too, but compressor blades are often provided with a special coating to improve their corrosion and erosion resistance.

5.3 COMBUSTOR

The combustion liner is exposed to much higher temperatures that the compressor components and neither titanium nor iron-based alloys exhibit suitable resistance to the various types of heat induced failure that can occur. Instead, the most common type of material for combustor liners has been nickel superalloys. Hastelloy X was used between the 1960s and the 1980s when it was replaced by an alloy called Nimonic 263. However, even nickel alloys could not cope with the subsequent increase in combustion temperatures and cobalt based superalloys are now replacing them in the most demanding situations. Where temperatures exceed even the ability of these alloys, special ceramic coatings with low heat conductivity are added to help with thermal resistance. This is combined with air cooling of the metal components to keep the temperatures within the operating range of the superalloys. The same materials and coatings may be used for transition pieces too.

5.4 TURBINE COMPONENTS

The most demanding conditions within a gas turbine are normally found in the first stages of the power turbines and it is here that the

[2]Materials for Gas Turbines—An Overview, Nageswara Rao Muktinutalapati, Advances in Gas Turbine Technology, Intech, 2011.

Table 5.1 The Composition of Alloy 718	
Element	Proportion
Nickel	53.0
Chromium	19.0
Iron	18.5
Molybdenum	3.0
Titanium	0.9
Aluminum	0.5
Cobalt	5.1
Carbon	(0.03)
Source: *Intech.*	

most advanced materials technologies are applied. Iron alloys have been used in the past to fabricate both stationary and rotating components but as temperatures increased, these were replaced by a range of nickel-iron and nickel-based superalloys. Typical of these advanced materials is an alloy denoted by the title Alloy 718 which is used for large rotating power turbine components in large industrial gas turbines. It is made from nickel, chromium, iron, molybdenum, titanium, aluminum, cobalt, and carbon. The composition is shown in Table 5.1.[3] Even these nickel-iron and nickel superalloys have not proved adequate for the latest power turbine inlet stages and they are now being replaced by cobalt superalloys for the most demanding situations.

These most demanding conditions are met in the first stage vanes and blades of the power turbine. The vanes are stationary but they are the first components that the gases from the combustor reach and they must be able to withstand the highest temperatures, as well as being resistant to corrosion from any impurities carried through with the hot, high pressure gases. The first stage turbine blades are also exposed to extremely high temperatures. However, in their case the difficulty is compounded by extremely high rotational speeds that produces massive centrifugal forces within the blades themselves, forces that can tear the component apart. The forces manifest themselves through what is known as creep rupture of the metal component. Creep rupture is specific to high temperature conditions and can lead to component

[3]Materials for Gas Turbines—An Overview, Nageswara Rao Muktinutalapati, Advances in Gas Turbine Technology, Intech, 2011.

fracture. It can be found in turbine blades and also in the discs that hold the blades in position on the turbine shaft since these have to support all the centrifugal force generated by the mass of the blades.

In order to create materials that can operate under such conditions, not only have new materials been developed but new methods of preparing them have had to be introduced. The traditional method of manufacturing metallic components is to forge them by machining from ingots of the material. Alloys designed to be forged have to exhibit specific properties compatible with forging. As inlet temperatures have risen it has proved difficult to combine the properties that lead to ease of forging with the properties required to retain high temperature strength and corrosion resistance. To get around this, there was a shift to the use of investment casting[4] to form the components precisely without the need for forging.

Casting using a mold and molten metal produces a cast component which has a microcrystalline structure with randomly sized and oriented grains. The rate at which the cast is cooled can be used to control grain size to some extent. These cast components enabled higher operating temperatures to be reached but eventually even they reached their limit.

The main failure mechanism in turbine blades made from such castings involves cavities forming along grain boundaries that occur perpendicular to the length of the blade and to the direction in which the centrifugal force is acting. To overcome this, a technique called directionally solidified casting was developed which preferentially reduced the number of these transverse grain boundaries by extending the grains along the length of the blade. This helped increase the temperature capability of such castings by around 14°C compared to a conventionally cast component.

Directionally solidified cast components remain adequate in some extreme situations but in others even these were not capable of providing adequate lifetimes under the first stage power turbine conditions. The solution was single crystal casting. In this type of casting a component such as a turbine blade is cast as a single crystal of the superalloy.

---

[4]Investment casting involves the lost "wax process" in which a model of the component is made in wax and then the model is encased in a ceramic to create a mold. Hot metal is then poured into the mold where it displaces the wax and forms an identical component.

Further, the orientation of the crystal can be controlled so that the direction of greatest strength is exactly that which will experience the greatest tension and stress.

There are additional advantages of single crystal component growth. As these contain no grain boundaries, elements that are added to the superalloy mix to control strength at grain boundaries can be eliminated and this has led to higher melting point alloys, improving further their high temperature strength. Even so these single crystal castings have their limitations; even they begin to soften at the power turbine inlet temperatures. To overcome this, special coatings have been designed that can protect the alloy surface from the most extreme temperatures.

5.5 COMPONENT COATINGS

One of the main ways of improving the performance of superalloys in gas turbines is with coatings. These coatings can be used to increase the resistance to corrosion or oxidation of the metal surface. Other types of coating, when used in conjunction with a cooling regime, can help protect the component from the most extreme temperatures.

Superalloys for high temperature components have to combine high physical strength such as resistance to creep fatigue and erosion with resistance to chemical erosion of the metal surface via either corrosion or oxidation. However, as the conditions that the metals must resist become more and more extreme, so it becomes more difficult to combine all these properties in a single superalloy. Coatings offer a way of enhancing the superalloy performance.

There are two types of environmental degradation to which hot turbine components can be subjected, hot corrosion and high temperature oxidation. High temperature corrosion occurs when there are alkali metal contaminants such as sodium and potassium present as well as sulfur. Sodium can be found, for example, if the plant is operating near the sea and there is salt water spray in the air. Sulfur usually comes from the fuel. Natural gas contains little sulfur but many natural gas plants are also designed to operate on liquid fuel too and this may contain some sulfur. Meanwhile oxidation is a natural process, the reaction between a metal and oxygen from air, that takes place more rapidly the higher the temperature.

In order to combat high temperature corrosion and oxidation, the normal strategy is to add elements to the alloy that help resist the corrosive reaction. However, these can have adverse effects on the high temperature performance of the alloy.

Coating the surface of the superalloy can help prevent or slow these phenomena. The most common type of coating is a diffusion coating, usually of an aluminum compound called an aluminide. This is formed by depositing a thin layer of aluminum onto the surface of a component made of a nickel or cobalt superalloy and heating it to around 1000°C. At this temperature some of the aluminum diffuses into the surface of the alloy to create cobalt-aluminum or nickel-aluminum compounds called aluminides which have higher corrosion and oxidation resistance than the superalloy alone. At the same time the aluminum will react with oxygen at high temperature to preferentially form aluminum oxide which also helps form a barrier coating. If this oxide layer becomes damaged and eventually flakes off, a new layer forms provided there is still aluminum present to form it.

These aluminide coatings rely on the substrate superalloy providing a part of the coating. An alternative is to apply an overlay coating that does not require reaction with the substrate. This type of coating can contain a variety of components that are tailored to provide the necessary resistance. In addition the thickness of the coating does not depend on the ability of the diffusion coating to react with the substrate. Overlay coatings can be laid down using a vacuum plasma spray method. A typical overlay coating will contain nickel (or cobalt), chromium, aluminum and yttrium.

5.6 THERMAL BARRIER COATINGS

While the coatings discussed above are primarily designed to increase resistance to corrosion or oxidation there is another type of coating that is designed to improve the ability of the component to withstand higher temperatures. This type of coating is called a thermal barrier coating. Thermal barrier coatings are usually applied on top of an overlay or diffusion coating.

A thermal barrier coating is a thin layer of a material that has a very low thermal conductivity. When it is applied to a turbine blade it creates a thermal barrier layer between the hot gases from the gas turbine combustor and the superalloy blade itself. The thin layer will conduct heat only slowly. On its own, this confers no great advantage because eventually the heat will be transmitted to the component underneath and the latter will reach the external temperature. However, if the blade is also cooled internally, then the metal material can be kept within the temperature range at which it can operate while the barrier coating withstands the external temperature. These thin barrier coatings can support a temperature difference of hundreds of degrees centigrade across them.

The superalloys that are used to make turbine blades start to soften in the 1200°C to 1400°C temperature range. If a blade is coated with thin layer of a material such as zirconium oxide or yttrium oxide then the temperature at the metal surface can be kept below the softening temperature. In order to do this the thermal barrier coating must be able to support a temperature drop of 200°C or more across a very thin film.

Maintaining the integrity of the thermal barrier coating can be extremely difficult. It requires that the coefficient of expansion of the superalloy is closely matched to the barrier coating. Otherwise one will expand at a different rate to the other and the two will separate. The intermediate overlay coating for corrosion resistance may be used to help match the expansion coefficients across the surface layers. Thermal barrier coatings are used both on power turbine components and also on the inner surface of the combustor liner. The latter is cooled externally with air from the compressor.

5.7 TURBINE BLADE COOLING

In order for a thermal barrier coating to be effective, a turbine blade must be cooled internally so that the temperature of the underlying superalloy can be maintained below the point at which the metal will start to soften. Internal cooling involves creating cooling ducting and channels within the shaft or discs that support the blades to feed the cooling fluid to the hollowed blades. Depending upon the design there will also be holes and slots in the blades themselves to allow the cooling fluid to pass through them and out. This is shown schematically in Fig. 5.1.

The simplest form of cooling involves stealing some hot air from later stages of the power turbine or from the compressor and using this to cool the early stages of the power turbine. The cooling air enters the

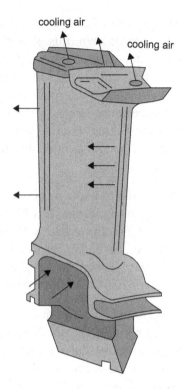

cooling air

cooling air

Figure 5.1 Turbine blade cooling. Source: From Paul Breeze, Power generation Technologies, 2nd Edition Newnes 2014.

blade through its root and is vented into the gas turbine through holes in the blade. This air then forms a film across the surface of the blade which protects the blade material, hence its name, film cooling. Eventually this cooling air becomes mixed with the flow of air through the machine. Borrowing air in this way will reduce the efficiency slightly but offers the simplest way of providing cooling.

The alternative is steam cooling. This can be applied when the gas turbine forms part of an integrated combined cycle system with a steam generator. Steam is an effective cooling fluid and it carries a smaller efficiency penalty than using borrowed air. However, it is much more difficult to implement because the steam cooling circuit must be kept apart from the air passing through the turbine. Its use can also have an impact on the flexibility of the overall plant. As a result of the complexity, some companies use steam cooling for stationary components only and air cooling for rotating components. Steam cooling can also be applied effectively in the combustor.

5.8 ADVANCED MATERIALS

Of an increase of roughly 500°C in power turbine inlet temperature over the past six decades it is estimated that around 150°C of the advance is due to the introduction of new superalloys and use of directionally solidified and single crystal fabrication techniques while a further 100°C can be attributed to the use of thermal barrier coatings. However, existing materials are now reaching the limit of their applicability and new ones are needed to enable temperatures to rise further. Candidates include using alternative metals as the basis for alloys and the use of ceramics.

Among the alternative metals that might be used are chromium, molybdenum and platinum. Chromium-based alloys are known to have higher melting points, good oxidation resistance and lower density that nickel-based alloys as well as thermal conductivity that is up to four times greater than many superalloys. However, chromium is extremely brittle at higher temperatures and this is exacerbated if it is exposed to nitrogen which is present in large quantities in air. This has so far prevented the development of chromium superalloys that might be suitable for gas turbines.

Molybdenum alloys also have a very high melting point and can be used in very high temperature applications. Unfortunately the metal oxidizes very rapidly above 500°C so the alloys can only be used in an oxygen-free atmosphere. Platinum-based alloys are attractive because of their stability at high temperatures, resistance to oxidation, ductility and thermal conductivity. Price is the main handicap that prevents more widespread use but it is envisaged that platinum superalloys may find application in some extremely demanding stationary components in gas turbines.

The other alternative is to use ceramic materials. Silicon carbide and silicon nitride have been considered as potential candidates since the 1960s. They can operate at much higher temperatures than metal alloys, they are resistant to corrosion and oxidation, they are much lighter than high temperature superalloys and they are much cheaper. However, all ceramics so far tested have been far too brittle to be useful. Until this problem can be overcome, it is unlikely they will find widespread application in gas turbines.

CHAPTER 6

Advanced Gas Turbine Cycles

The basic gas turbine as outlined in the preceding chapters comprises three sections, a compressor, a combustion chamber and a power turbine. The compressor—as its name implies—compresses inlet air. This is then fed into the combustion chamber where fuel is added and ignited, adding energy which increases both temperature and pressure of the gas. The hot, high pressure as is then allowed to expand through the power turbine to generate shaft power to drive a generator.

There are a number of modifications that can be made to this simple cycle in order to try and improve its overall efficiency. One of simplest is called recuperation. This involves the use of heat from the gas turbine exhaust to heat the air from the compressor before it enters the combustion chamber. A second strategy is called reheating. In a reheat system the power turbine is split into two parts, a high pressure turbine and a low pressure turbine. Hot gases exiting the high pressure turbine are further heated in a second combustor before entering the low pressure turbine. A mirror image of this is intercooling, which is applied to the compressor by dividing it into two parts and cooling the air between the two. A final method of increasing overall efficiency is called mass injection. This usually involves injecting water or steam into the air at some stage during the gas turbine cycle.

6.1 RECUPERATION

A simple gas turbine is a relatively efficient machine but it does waste a significant amount of the heat energy generated in its combustion chamber. This heat energy is carried away in the exhaust gases from the turbine. The exhaust gas temperature of a typical simple cycle gas turbine is in the range 400−500°C. If some of this heat energy can be used in the gas turbine cycle it will improve the overall cycle efficiency. This is what recuperation can achieve.

Gas-Turbine Power Generation. DOI: http://dx.doi.org/10.1016/B978-0-12-804005-8.00006-9

In a recuperated cycle the gas from the power turbine exhaust is passed through a device known as a counter-current heat exchanger, usually called more simply a regenerator or recuperator, before it is released into the atmosphere. Passing through the recuperator in the opposite direction is the air that exits the compressor. So long as the exhaust gases are hotter than the air from the compressor, heat will be passed from the exhaust to the air from the compressor before it is fed into the combustion chamber. By raising the temperature of the input air, the amount of energy needed to achieve the required temperature of the gases at the combustor outlet is reduced, cutting overall fuel consumption. A recuperated gas turbine cycle is shown in Fig. 6.1.

There are some serious limitations on the application of recuperators to gas turbines. First, the gas exiting the compressor must have a lower temperature than the gas exiting the turbine exhaust. With high compression ratio compressors, the exit gas temperature can be as high as that of the exhaust gas so the complete cycle must be designed around the use of recuperation.

Secondly the effectiveness of the recuperator depends on the amount of heat it can transfer from the exhaust gas to the input gas. The efficiency of a heat exchanger usually depends upon its surface area; the larger the surface area to which the two gas flows are exposed, the greater the heat transfer. However, the heat exchanger will have a resistance that will

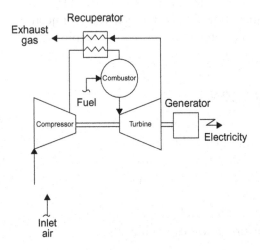

Figure 6.1 A gas turbine with recuperation. Source: With permission from Pinchco bvba.

reduce the pressure of the air from the compressor. The larger the heat exchanger, the larger the pressure drop. Ideally any pressure drop should be kept below 2% and this will limit the size of the heat exchanger that can be used. Large, efficient heat exchangers are also costly.

In consequence, commercial recuperated gas turbines are relatively rare. One such is a machine that was developed during the 1990s as part of the US Department of Energy's Advanced Turbine Systems program. This led to a 4.2 MW gas turbine being manufactured by US company Solar Turbines.[1] The machine had a compression ratio of 9.1:1 and a power turbine inlet temperature of 1163°C. The first stage turbine blades used single crystal alloy. Efficiency was quoted as 40.2%.

While industrial turbines with recuperators are rare there is one area in which they are becoming more common, microturbines. These tiny gas turbines are designed to supply single dwellings or small commercial organizations. Microturbines will be discussed in more detail in Chapter 8.

A much more complex recuperation cycle that has attracted some attention in recent years is called the chemically recuperated gas turbine. This uses the exhaust gas heat from the gas turbine to raise steam which is used to drive a chemical reaction which converts (reforms) a fuel, such as diesel, into hydrogen which can then be burnt in the combustion chamber of the turbine. This configuration is claimed to be able to increase efficiency and reduce the emissions of nitrogen oxides significantly. However it has yet to be applied to a commercial gas turbine.

6.2 REHEATING

Another strategy aimed at increasing the efficiency of a gas turbine is called reheating. This technique is common with steam turbines in large fossil fuel power stations but is less common in gas turbine design.

Many large fossil fuel steam power plants use not one but several individual steam turbines in order to capture the most energy from the steam working fluid. A typical arrangement would involve a single

[1]Current power output is 4.6 MW.

high pressure turbine, a single intermediate pressure turbine with larger blades and then two low pressure turbines with very large blades. The sizes and speeds of the turbines are optimized for energy capture at the different steam temperature and pressure ranges at each stage. In this type of plant overall efficiency can be raised by taking the steam that exits the high pressure turbine and heating it further to add energy before feeding it into the intermediate pressure turbine. The adding of energy at this stage is called reheating.

Gas turbine plants do not usually involve multiple turbines in the same way as a steam plant might. However it is possible to divide the power turbine of a gas turbine into two sections or spools.[2] This is done, for example, in some aero derivative gas turbines with one spool driving the compressor and the second driving the generator. There is no reason, in principle, why this division of the power turbine cannot be used to build a reheat gas turbine and one of the major gas turbine companies has achieved success with its heavy duty gas turbines by doing just that.[3]

In a reheat gas turbine the hot, high pressure gases from the combustor enter the high pressure spool of the gas turbine where energy is extracted and the temperature and pressure drops. The cooler gases exiting the high pressure spool then enter a second combustor where additional fuel is burnt to raise the temperature of the gases again. These gases then enter the second turbine spool where, because the temperature has been increased, more energy can be extracted than would have been possible at the lower temperature at which they exited the first spool. The gas turbine reheat cycle is shown in Fig. 6.2.

The French company Alstom successfully incorporated this design feature into its heavy duty gas turbines. The design used a higher compression ratio, 30:1, than most other industrial gas turbine designs. The use of reheat allowed the inlet temperature at the first stage of the high pressure power turbine spool to be lower than in comparable turbines without reheat and so permitted less costly materials and cooling designs to be used. The design had significant efficiency advantages when operating in open cycle at part load, a feature that will be

---

[2]A spool is a set of turbine or compressor blades that all rotate together and at the same speed.

[3]Alstom has two turbine ranges that include reheat. The gas turbine part of the French company was acquired by GE in 2014 and the fate of these unique turbines is unknown.

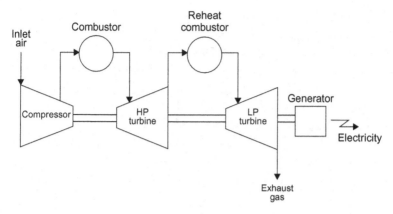

Figure 6.2 A gas turbine with reheat. Source: With permission from Pinchco bvba.

increasingly important for future gas turbine power plants. So far no other company has emulated this design for industrial gas turbines.

6.3 INTERCOOLING

If reheat requires the power turbine of a gas turbine to be divided into two spools, intercooling does the same for the compressor. Instead of a single set of stages, the compressor is divided into two sets, or spools. The compressed air from the first spool is then passed through a cooler—the intercooler—before entering the second spool.

The intercooler is a heat exchanger that extracts heat from the compressed air. It can be an air to air heat exchanger where ambient air is the cooling fluid or, more commonly, a water to air heat exchanger that uses cold water to reduce the temperature of the compressed air from the first spool of the compressor. A gas turbine with intercooling of the compressor is shown schematically in Fig. 6.3.

The advantage of intercooling is that it reduces the volume of the compressed air and so reduces the amount of work that the compressor has to do in order to achieve the required pressure at its outlet. This means that it will consume less of the energy produced by the power turbine, leaving more for power generation. It also allows a higher pressure to be achieved with the compressor. The result is an overall increase in engine efficiency.

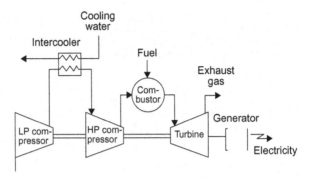

Figure 6.3 A gas turbine with compressor intercooling. Source: With permission from Pinchco bvba.

Other similar strategies include cooling the inlet air with a refrigeration system, or by spraying a mist of water into the intake. Like intercooling, this is effective because it reduces the amount of work the compressor has to do to reach a specified compression ratio. However the addition of water to the air has other effects too, as are discussed below.

Intercooling has been implemented in an aero derivative turbine manufactured by GE called the LMS100. This machine has a six stage low pressure compressor spool. The air exiting this spool is then taken into an off-unit intercooler that can be either air or water cooled. After being cooled the air is returned to the high pressure compressor spool. With intercooling, an overall compression ratio of 42:1 is achieved. The compressed air then enters the combustor where the temperature is raised to 1380°C before entering the first of three power turbine spools. The first, two stage high pressure turbine is used to drive the high pressure compressor spool. This is followed by a two stage intermediate pressure turbine which drives the low pressure compressor. Finally a five stage power turbine drives the generator. The system can generate between 100 and 110 MW depending on the precise configuration and has an efficiency of 46%.

6.4 MASS INJECTION

A final strategy for increasing the efficiency of a gas turbine is called mass injection. This involves introducing water or steam at some stage during the gas turbine cycle. Steam may be injected into the combustor of a gas turbine where it can also be used to lower the firing temperature and reduce the production of nitrogen oxides. This will at the same time result in an increase in power output. An increase in performance can also be achieved by injecting water into the compressor. The idea was first explored during the early 1970s and Dah Yu Cheng obtained a number of patents covering this type of cycle.

The use of mass injection is a strategy to get around one of the basic efficiency problems in a gas turbine. The high combustion temperature in the combustor, peaking at around 1900°C with natural gas if it is burnt in a stoichiometric amount of air, is too high for the components in the hot gas path. So, in order to reduce the temperature, excess air beyond what is needed for the combustion process is added. Excess air helps reduce the combustion temperature. However, this excess air must be compressed along with the air that will react with the natural gas and this additional compression represents a parasitic load that reduces the overall efficiency of the cycle.

Steam injection into the combustor of a gas turbine helps to cool the combustion gases and as a result less excess air is needed. In addition the steam adds to the mass passing through the gas turbine, increasing its power output. Adding steam in this way, or it other ways discussed below, is a little like combining a steam turbine and a gas turbine in a single cycle. This can be seen as a cheaper alternative to the combined cycle plant discussed below in Chapter 7. However, it is not as efficient as the latter.

The most common form of mass injection is the steam injected gas turbine cycle which is exploited in a number of commercial gas turbine systems. In this type of cycle the waste heat from the exhaust of the gas turbine is used to produce steam and this steam is then injected into the gas turbine, as shown in Fig. 6.4. Steam injection can take place either at the start of the combustor or between the combustor and the power turbine. Injection in the combustor is normally preferred because the steam will cool the combustor.

The addition of steam increases the mass of gas passing into the power turbine. This results in a higher power output from the turbine that would be achieved without steam input. The generation of the steam has an energy cost but this was energy that would have been wasted otherwise. The net result is an increase in efficiency compared to the same turbine without steam injection of between 2% and 4%.

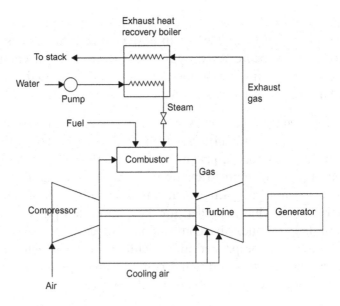

Figure 6.4 Schematic of a gas turbine with steam injection. Source: NASA Government.

A similar effect can be produced by introducing a fine spray of water into the inlet of the compressor of a gas turbine. This procedure, called inlet fogging, affects the performance of the gas turbine by cooling the inlet air and by an intercooling effect within the compressor, both of which help boost output of the machine.

More complex mass injection cycles are also possible. One of the most explored is the humid air turbine (HAT) cycle, shown in one version in Fig. 6.5. In this cycle the compressor is divided into two spools with an intercooler between them. There is also a recuperator that recovers heat from the turbine exhaust. The recuperator is used to heat the compressed gases just before they enter the combustion chamber, as in a conventional recuperated system. However, the lower grade heat from the recuperator, after it has been used to heat the gases entering the combustor, is also used to heat water. The heat from the intercooler is used for the same purpose. The heated water is used to humidify the compressed air that exits the compressor so that the air which finally enters the combustion chamber is saturated with water. This complex cycle is claimed to have a significantly higher efficiency

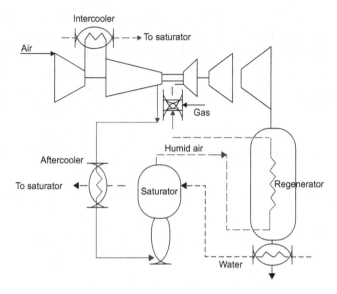

Figure 6.5 The HAT cycle. Source: From National Energy Technology Center, U.S. Department of Energy.

than a conventional gas turbine but without the cost involved in building a more complex combined cycle plant.

A further refinement of the HAT cycle, called the cascaded HAT cycle introduces a reheat stage, dividing the power turbine into two spools. This variation has a claimed efficiency of up to 55%. Other variants include the injection of water into the compressor stages and water injection at the compressor inlet. Yet another uses solar energy to add heat to the compressed gases before entry into the combustor. All these mass injection cycles release significant amounts of water vapor into the atmosphere and so require large volumes of water available to be able to operate. None of these cycles has yet been converted into a commercial gas turbine system.

6.5 COGENERATION

Cogeneration is another way of improving the efficiency of a gas turbine power plant. In this case the heat from the turbine exhaust, and from any other source in the cycle, can be used to generate either steam or hot water. The high temperature of a gas turbine exhaust makes this an ideal way of producing steam for plant that requires high temperature process heat. If that is not required then hot water can be produced for local space heating and hot water needs, or for a district heating system. An efficiently designed cogeneration system of this type can increase overall efficiency, when heat generation is included along with electricity generation, between 80% and 90%.

CHAPTER 7

Combined Cycle Power Plants

Adapting a simple cycle gas turbine in the ways outlined in Chapter 6 offers a means of increasing the efficiency of the cycle by reducing parasitic loads such as the excess air compressor load and of capturing energy that might otherwise be wasted. However, this is not always the most effective way of increasing overall efficiency of energy conversion. For large gas turbine based plants in particular, the best way of improving efficiency is to add a steam turbine bottoming cycle, creating a combined cycle power plant.

A combined cycle plant is simply what its name suggests. Instead of relying on a single thermodynamic cycle to convert energy into electricity the plant uses more than one. These piggy-back one another with the first cycle using the highest temperature thermodynamic working fluid, followed by a second using the intermediate temperature fluid and—in principle at least—a third using a lower temperature working fluid. In fact combined cycle plants with more than two cycles are not used commercially although they are theoretically possible.

One way in which a gas turbine can be used in this way is as a topping cycle. There are a number of power plant concepts that involve the generation of a high pressure, high temperature gas stream. A pressurized fluidized bed combustion plant, a high pressure molten carbonate fuel cell and some high temperature solar power plants can all produce a stream of gas that could be used to drive a gas turbine before the hot gas is used either to generate steam to drive a steam turbine, or in the case of the fuel cell, to provide fuel for the fuel cell. Topping cycles are not the subject of this book but more can be found about these individual topping cycle plants in other books in this series.

Much more significant for the modern power generation industry is the addition of a bottoming cycle to a gas turbine power plant. In this

Gas-Turbine Power Generation. DOI: http://dx.doi.org/10.1016/B978-0-12-804005-8.00007-0

case the bottoming cycle is usually a steam turbine cycle, with heat from the gas turbine exhaust exploited to raise steam. This is the most common combined cycle power plant.

It would be possible to add a third cycle to exploit the low grade heat remaining after steam generation. This could be achieved with a closed cycle turbine such as an organic Rankine cycle. Such turbines can exploit low grade heat to produce electricity and are used in some geothermal plants where the temperature of the geothermal reservoir is relatively low. However, it is unlikely to be economically viable to add this third cycle to a modern combined cycle plant.

7.1 EVOLUTION OF THE COMBINED CYCLE CONCEPT

The first gas turbine that was ever sold commercially for use in a power station to generate electricity operated in a sort of combined cycle mode. The 3.5 MW unit was installed at the Belle Isle Station in Oklahoma in 1949. Belle Isle was an established fossil fuel-fired steam power station and the heat in the exhaust gases of the gas turbine was used to heat the feed-water for the boiler of the steam plant, boosting steam plant output. In later applications of gas turbines, during the 1950s and 1960s, the exhaust gas from a gas turbine was used as the combustion air for a conventional fossil fuel-fired boiler. This added a few percentage points to overall efficiency. Other early combined cycle power plants were created by repowering fossil fuel plants, replacing the coal furnace with a gas turbine but using the steam generator and steam turbine of the original plant to exploit the heat in the gas turbine exhaust.

The development in the 1950s of more advanced boiler tubes with spiral fins welded to the central tube gradually made it possible to build a more effective heat recovery steam generator (HRSG) for energy capture from a gas turbine exhaust. Many of the first plants to employ this new technology were cogeneration plants that produced electricity and steam but a small number of these steam generators were used in utility power plants. This established the modern gas turbine power station.

The basic combined cycle configuration consists of a gas turbine, a HRSG and a steam turbine. In the early incarnations of this configuration the gas turbine operated as if it were a stand alone gas turbine generating

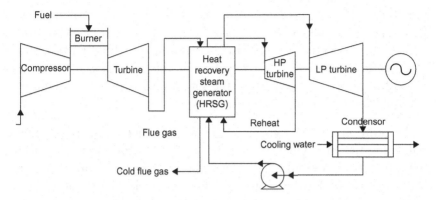

Figure 7.1 Schematic of a combined cycle power plant. Source: With permission from Pinchco bvba.

plant with its own generator producing electrical power. The steam bottoming cycle is then added. As we have seen in earlier chapters the temperature of the exhaust gases exiting a gas turbine is relatively high, usually 400°C to 500°C, and sometimes higher. This is hot enough for the heat to be captured in a HRSG where it is used to raise steam and this steam is then exploited in a steam turbine generator which produces more electricity. The basic combined cycle configuration is shown in Fig. 7.1.

Plants based on this configuration became common during the 1970s and 1980s. However, these were still based on a standard open cycle industrial gas turbine. Change came towards the end of the 1980s when the major manufacturers began to explore the idea of designing both gas and steam turbines specifically for combined cycle power stations. This eventually led to the modern generation of extremely high efficiency combined cycle plants. In 1990 the best efficiency for a combined cycle plant was 50%. In 2011 a combined cycle power plant in Germany achieved 60.75% efficiency.

7.2 THE HIGH EFFICIENCY COMBINED CYCLE POWER PLANT

The great advances achieved in combined cycle efficiency can be attributed to two principle sources. The first is the tight integration of all the components of the power plant. This allows the two cycles to operate at optimum efficiency and reduces energy loss. The second has been the advance in efficiency of the stand alone gas turbines as a result of the increase in first stage temperature that has been achieved with new materials and designs. The current designs in which the highest inlet temperature can reach 1600°C have broken the 60% efficiency barrier. Efforts are underway to push this temperature to 1700°C, potentially leading to an efficiency of 65%. To put this in perspective, the best coal-fired power stations using the most advanced boiler technology can only achieve around 47% efficiency.

Integration of the two cycles in a combined cycle plant means balancing the energy capture in each cycle. A stand alone gas turbine will offer its highest efficiency when the exhaust gas temperature is as low as possible. However, for a combined cycle plant it is advantageous to allow the exhaust gases from the gas turbine to leave at a significantly higher temperature since this makes the steam turbine cycle more efficient. So while the gases may leave a stand-alone gas turbine at a temperature as low as 400°C it is not uncommon to see the exhaust gas temperature for an integrated combined cycle unit to be over 600°C.

Much of the design effort that led to these high efficiency systems has been aimed at achieving high turbine inlet temperatures. Heavy duty industrial gas turbines have traditionally been given a letter of the alphabet that denotes the series to which they belong. Later letters in the alphabet denote a turbine with a higher inlet temperature. During the 1990s the biggest industrial turbines were F class turbines. H class gas turbines began to appear at the end of the 1990s and by 2011 there was at least one J class turbine.

The H class turbines generally have a turbine inlet temperature of between 1400°C and 1500°C. To manage this, companies have had to resort to extremely advanced turbine materials and blade cooling configurations. The first company to announce an H class turbine was GE which developed the machine under the US Department of Energy's advanced turbine system program. Inlet temperature is 1430°C and the compression ratio is 23:1, higher than the 15:1 ratio in the company's earlier F class machines. In the H turbine the first stage vanes and turbine blades both use single crystal superalloy with a thermal barrrier coating. In addition, the close integration with the steam turbine cycle in the H machine means that steam from the steam generator can be used to cool the vanes and blades of the first two turbine stages. The third stage uses more conventional air cooling and the fourth stage is uncooled.

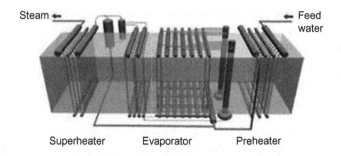

Steam ←   ← Feed
                                                        water

Superheater   Evaporator   Preheater

Figure 7.2 A heat recovery steam generator. Source: With permission from Nooter/Eriksen.

There is a complex steam cycle too. The HRSG provides steam at three pressure levels. The highest pressure steam is used in the first, high pressure stage of a two-stage steam turbine. Steam is then reheated before entering the second, intermediate stage steam turbine. The third pressure level is used for steam cooling. A typical HRSG layout is shown in Fig. 7.2.

Steam cooling offers an efficiency advantage over the more traditional air cooling because the latter usually borrows air from the compressor and so represents a further parasitic load. Steam from the combined cycle steam generator does not carry the same penalty. However, steam cooling is more complex to implement and it also affects overall flexibility. So while GE and the Japanese company Mitsubishi Hitachi Power Systems have both adopted steam cooling, Siemens has exclusively used air cooling for its high efficiency combined cycle plants. Alstom too, retained air cooling in its large gas turbines which are novel for their use of gas turbine reheat to avoid the high inlet temperatures used by other manufacturers. (Alstom sold its gas turbine division to GE in 2014, reducing the number of major manufacturers of large industrial gas turbines to four including Ansaldo.)

7.3 HEAT RECOVERY STEAM GENERATORS

The HRSG is a key component of a combined cycle power plant. Its role is to convert as much of the heat as possible from the exhaust gas of the gas turbine into steam for a steam turbine. The temperature of the exhaust gases from the steam turbine will be between 400°C and perhaps 650°C, low compared to the gases exiting the boiler of a coal-fired power plant. The heat in this gas is captured by water and steam that flows through tubes placed in the path of the hot gases. These tubes will have fins welded to them to increase their surface area so that they can absorb more heat from the hot gases. These tubes are arranged in modules, sometimes called racks, with each module serving a slightly different function.

From the perspective of water heating, the first module is called an economizer. This takes relatively low grade heat and uses it to heat the feed-water that is returning from the steam turbine to the boiler. This hot water is then fed into a module called the evaporator which heats the water to its boiling point. The temperature at which the water boils may be much higher than 100°C because the system is operated under pressure. At the top of the evaporator is a steam drum where water and steam are separated. The water cycles back through the evaporator while steam is collected a taken to the third module, called the superheater. This dries the steam and raises its temperature above the boiling point, before piping it to the steam turbine.

The three modules are arranged in the hot gas path so that the superheater is exposed to the hottest gas. This is followed by the evaporator while the economizer takes heat from the coolest exhaust gas, before it is released through the plant stack. Complexity can be added because many modern combined cycle plants use HRSGs with two or three of these economizer—evaporator and drum—superheater arrangements to provide steam at two or three pressures. In this case the multiple modules are arranged in order of decreasing steam and water temperature. In some designs there may also be one further module, called a reheater, which reheats steam from a high pressure steam turbine before it is fed to an intermediate pressure steam turbine as in the GE H system described above.

There are a number of design for HRSGs although they all share many common features. The variants include steam generators in which the exhaust gas path is vertical and the boiler tubes are arranged horizontally in the gas path and steam generators with a horizontal gas path and vertical steam tube modules. The way in which water and steam circulates within the boiler tubes varies too, with some using forced or pumped circulation and others using natural circulation. Natural circulation HRSGs usually have a horizontal exhaust gas path while forced circulation boilers are often vertical.

The temperature of the turbine exhaust gases fall as they pass through the various stages of the HRSG. Emission control to remove nitrogen oxides requires the gases to be at a specific temperature for the process to be carried out efficiently. The point at which the exhaust gases reach this temperature often lies within the HRSG. This means that these emission controls systems must be installed inside the boiler.

It is possible to add heat to the exhaust gases by installing gas burners within the steam generator, a technique known as supplementary firing. Supplementary firing can make the system more flexible by allowing more steam generation where needed or to supply additional heat if the gas turbines are operating at less than full load. Supplementary firing is less thermally efficient and is rarely used in very large combined cycle plants. However, it is common in stations that supply steam for process heat as well as electricity.

An advanced HRSG design that is being introduced to add greater operational flexibility to combined cycle power plants is the once-through steam generator. Fig. 7.3 compares the drum-type and once-through steam generator. This design eliminates the steam drum from the conventional HRSG design by completing the conversion from water to steam within the evaporator sections of the boiler. The steam drum of the more conventional design is a handicap for fast start-up because it is massive

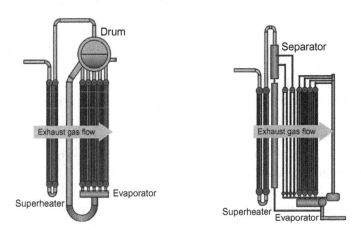

Figure 7.3 Schematics of a drum-type steam generator (left) and a once-through steam generator (right).
Source: With permission from Siemens.

and therefore takes a long time to reach its operating temperature. Eliminating it allows the steam cycle to start more quickly. Once-through steam generators can be both horizontal and vertical.

7.4 FLEXIBLE COMBINED CYCLE POWER PLANTS

The development of combined cycle gas turbine power plants has led to impressive gains in performance and, at over 60% energy conversion efficiency, the best modern stations offer the highest energy conversion efficiency of any large scale power plant. This equates to lower overall emissions of carbon dioxide for each unit of electricity generated.

These impressive performance figures can only be achieved under relatively steady state operating conditions. However, some of the strongest markets for large combined cycle plants are in developed countries that have sophisticated grid systems where large quantities of renewable electricity are being generated and delivered into the grid. The main types of renewable generation in use, wind power and solar power, are both intermittent and can display high levels of unpredictability. To manage this type of power delivery while maintaining a stable electricity network that can balance demand and supply requires a fleet of power stations that can take up the load when renewable input falls and back out when it rises. In many cases new combined cycle power plants are required to perform this function. This has led to the need to develop flexible combined cycle power plants.

The demands on a flexible plant are different to those experienced by one that operates under relatively stable conditions. A plant that is supporting renewable generation must be able to start-up quickly and be able to change its output level quickly too. This involves rapid changes in temperature which can create large temperature gradients in components. Such gradients are a major source of stress and fatigue leading to high maintenance and repair costs. Simplification of designs, such as the reduction of thermal mass and inertia in the once-through steam generator, can help reduce these stresses by allowing components to heat up more quickly. Other changes to the operation of combined cycle plants are intended to reduce the temperature swings.

One way of improving the ability of a plant to react quickly is to decouple the gas and steam turbine cycles. Gas turbines can start rapidly but the HRSG and steam turbine sections may require more time to reach operating temperature. Decoupling leads to complications for steam cooled gas turbines because the gas turbine cannot operate until the HRSG is operating at its correct temperature. Manufacturers have had to adapt. For example, GE has now developed a version of its H class gas turbine that is fully air-cooled,[1] so avoiding this problem. The new version offers faster start-up and ramp rates than the earlier steam cooled versions of the same turbine.

The period of greatest thermal stress in a combined cycle power plant is during a cold start after the temperature of all the components has been allowed to fall close to ambient. One way of avoiding this is to keep the plant warm by idling at very low power. "Parking" a power plant in this way not only avoids the stresses of a cold start but it means that the plant can be brought up to full power much more quickly. There is an economic price since the plant will burn fuel while it is parked but this may be more than offset by the savings in maintenance costs due to the wear and stress introduced by frequent cold starts.

Another issue that needs to be addressed in flexible combined cycle plants is the efficiency at part load. While a combined cycle plant may be able to achieve high efficiency at full load this will often fall off at part load as the operating conditions change. It may be necessary to design slightly lower full load efficiency in order to get high efficiency at less than full load. Typical part load efficiency for a flexible combined cycle plant will be around 50% to 55% under low load compared to 58−59% at full load.

The other area that is affected by flexible operation is power plant emissions. Emissions can be controlled relatively easily under steady state conditions but when conditions are varying, then emissions can rise significantly. (Think of a diesel-engined car accelerating and producing a cloud of black smoke.) If efficiency drops then the amount of carbon dioxide produced for each unit of electricity rises. Nonsteady state operation will usually increase the quantity of NO_x generated in the turbine combustor too. The latter can usually be removed using advanced emission control technology but it can still

---

[1]According to GE the new air-cooled versions of its H class turbines have open cycle efficiency of 41.5% and a combined cycle efficiency of over 61%.

lead to an overall rise in emissions compared to the steady state and varying conditions make control of emissions more difficult to maintain. These are problems that gas turbine and combined cycle plant manufacturers are addressing in the middle of the second decade of the 21st century.

7.5 INTEGRATED SOLAR COMBINED CYCLE POWER PLANT

The combined cycle power plant is a flexible concept and it can be adapted in various ways to accommodate different sources of energy. One of the most interesting of these is the integrated solar combined cycle (ISCC) power plant. This type of plant collects solar heat energy and adds it to the energy from fuel burnt in a conventional combined cycle plant in order to reduce the cost of power.

Solar thermal energy can be harvested in various ways. In the case of the ISCC plant the usual arrangement is for solar energy to be collected using an array of parabolic trough solar collectors. These are aligned so that they can track the sun across the sky from morning to night, collecting as much solar energy as possible and focusing it onto a heat collecting tube that runs along the length of each trough at the focus of the parabola. A heat transfer fluid is pumped through these heat collection tubes. By passing the fluid through several parabolic trough collectors in succession, the temperature of the fluid can be raised as high as 550°C.

In the simplest ISCC configuration the high temperature heat transfer fluid is passed through a heat exchanger that forms part of the HRSG of a combined cycle plant, as shown in Fig. 7.4. The heat is extracted and used to raise steam for the steam cycle of the plant. Provided the solar input is kept low compared to the total energy input into the steam cycle, the efficiency is much higher than for a stand alone solar plant and improves the efficiency of the combined cycle plant. However, the solar power input is only available during the day so that at night the steam temperature and pressure will fall because less energy is available. Depending upon the amount of solar energy available, this can have a significant impact on overall efficiency. Ideally the solar input should be kept to 10% or less of the total energy input.

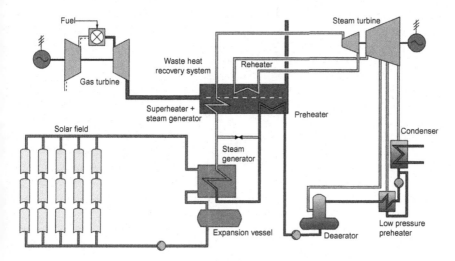

Figure 7.4 Schematic of an ISCC power plant. Source: With permission from Mr. Herrmann.

A way of using more solar energy and maintaining the steam operating conditions is to equip the plant with thermal energy storage. Various storage systems have been tested including water and molten salts. If thermal energy can be stored then it will be available over 24 hours instead of only being available during daylight hours.

An ISCC plant offers a very efficient way of using solar heat energy. It also offers an advantage in a flexible combined cycle plant because the solar energy can be used to maintain the steam turbine cycle at operating temperature, allowing much faster start-up when power from the plant is needed.

The first ISCC plant of this type was inaugurated in 2010 in Italy when a solar collection field was added to an existing combined cycle power plant near Syracuse in Sicily. Since then plants have been built in Florida, Egypt, Iran, Algeria and Morocco. This type of power plant is considered particularly attractive in regions such as North Africa where there is abundant sunlight and a ready supply of natural gas.

CHAPTER 8

Microturbines

Historically most of the gas turbines that have been used for power generation have been relatively large, with electrical outputs of over 1 MW. Many have outputs of several hundred megawatts. However there is a group of much smaller turbines, called microturbines. These turbines are intended for use in distributed generation applications where they supply electrical energy, and often heating or cooling too, to group of local energy users. For the smallest of these microturbines, the users might be in a single household. Even smaller units, called ultra-microturbines are being developed for portable use, as a replacement for batteries.

There is no standard definition of a microturbine. Some commentators include all turbines with an electrical output of less than 1 MW. This covers a broad field since the smallest of them can be of the order of tens of millimeters in diameter and have outputs of 10 to 100 W. For stationary applications the minimum size is likely to be 1 to 5 KW, sufficient to provide power for a small domestic dwelling. However, most of the commercial microturbines available are much larger with sizes ranging from 30 to 500 kW. These larger machines can also be deployed in parallel to provide even larger microturbine installations. The market for these devices is still evolving and the may take several more years to establish a firm position. Many commercial microturbine systems are still under development.

8.1 MICROTURBINE TECHNOLOGY

While the main gas turbine technology can be traced back to the development of aero engines during the middle years of the 20th century, microturbines are generally considered to have evolved from the auxiliary turbines that are fitted to aircraft to supply electricity and heat.[1] Similar devices were also tried as automotive power units in the 1970s but the technology did not thrive.

---

[1]Some may be related to turbochargers too.

Gas-Turbine Power Generation. DOI: http://dx.doi.org/10.1016/B978-0-12-804005-8.00008-2

Microturbines are small gas turbines and being small they are generally also much simpler than their larger relatives. A typical microturbine will use a radial compressor and turbine rather than the axial components that are common in larger gas turbines. There will probably be a single stage to the compressor and a single stage turbine too, both mounted on the same shaft, with a combustion chamber between the two. A high speed generator will share the shaft, making for a compact design. Air bearings are often used to minimize friction although conventional oil-lubricated bearings are also common.

These small turbines have extremely high rotational speeds, usually between 40,000 and 120,000 rev/minutes, and the output frequency of the generator is similarly high, perhaps as high as 1000 Hz. In order to match grid frequency, these systems are therefore fitted with power electronic frequency conversion hardware. This has the advantage of conditioning the supply and controlling both voltage and frequency and the unit should be able to operate independently of the grid if necessary.

The compression ratio that can be achieved using a single stage radial compressor in a microturbine is low compared to a conventional multistage compressor of larger gas turbines. Typical compression ratios are 3:1 to 4:1. In order to attain high efficiency these machines usually need to incorporate a recuperator. A microturbine with a recuperator is shown schematically in Fig. 8.1 while Fig. 8.2 shows a cross-section of a commercial microturbine. As discussed in Chapter 6 the recuperator captures energy from the exhaust gases of the turbine and uses it to preheat the compressed air from the compressor before it enters the combustion chamber. The typical efficiency of a microturbine with a recuperator is 25% to 30%. There are also microturbines that do not use recuperators. These devices tend to be more sturdy than those with recuperators but their efficiency is low at around 15%.

All the components of the microturbine will usually be integrated into a simple package that is ready to link to the mains supply. This ease of installation is one attraction of this type of device. Nevertheless, the efficiency that can be attained is low compared to many other forms of generation. This can be offset if further waste heat, in the exhaust gases exiting the recuperator, is used to heat water in a combined heat and power system. This will raise the overall efficiency to as high as 85% although it may still fall short of alternatives,

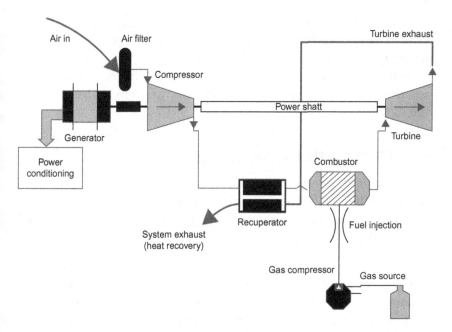

Figure 8.1 Schematic of a microturbine with recuperation and waste heat recovery. Source: With permission from National Institute of Building Sciences. Courtesy of EPRI.

Figure 8.2 Cross-section of a microturbine. Source: With permission from Capstone Turbine Corporation. Courtesy of Capstone.

such as a reciprocating gas engine. The waste heat can also be used to drive a chiller for cooling. Where cogeneration is important, a simple microturbine without a recuperator can be more effective since there is more waste heat energy available.

The area where microturbines may offer their greatest advantage is in their emissions performance. The main pollutant generated in a gas turbine is NO_x and the level of these emissions from a microturbine are usually very low. This will usually allow microturbines to be installed without the need for additional emission control systems, even in urban areas.

8.2 MICROTURBINE ENHANCEMENTS

Current microturbines have limited appeal as electricity generators because of their low energy conversion efficiency. Other limitations include a drop off in power as the ambient temperature rises and a fall in power at higher altitudes.

There are a number of enhancements that can improve the performance of a microturbine although not all will be cost-effective.[2] One that may be cost-effective is inlet air cooling. This has two advantages. The first is to insulate the turbine from changes in ambient air temperature. The second is to increase power output under normal operating conditions by reducing the inlet air temperature, so increasing the temperature difference between inlet and outlet. Air cooling can be carried out either by refrigeration of the air before it is delivered to the input or by inlet air fogging, as discussed in Chapter 6. The latter also adds an element of mass injection.

A second possibility is to create a micro-combined cycle plant by using a closed cycle organic Rankine turbine to exploit the heat from the microturbine exhaust. This is a complex and potentially costly solution to the efficiency problem but has the potential to raise the efficiency significantly. Another cycle modification is to apply mass injection of steam into the combustor of the microturbine. Again this adds complexity and may not be cost-effective.

Other ways of increasing the performance of microturbines involve the same strategy used in larger turbines, by increasing the inlet temperature at the turbine. The small size of a microturbine means that the production of high temperature ceramic components may be much more cost-effective than for larger machines. Thermal barrier

[2]Further details about these enhancements can be found in Micro Gas Turbines, Flavio Caresana, Gabriele Comodi, Leonardo Pelagalli and Sandro Vagni, in Gas Turbines, published in 2010 by Intech.

coatings, not normally used in microturbines, might also be introduced. As with other possible enhancements, cost is likely to be the determining factor.

8.3 APPLICATIONS FOR MICROTURBINES

There are, in principle, a wide range of applications for microturbines. One major use is likely to be for standby or backup power. This might be in a hospital or a data center where power availability is critical but it could also be in less sensitive commercial settings if the grid supply is insufficiently reliable. In addition, companies could use microturbines for peak shaving, using the unit to supply electricity when the cost from the grid is high but returning to the grid when costs fall. In regions where demand management is common grid practice, microturbines might offer small companies a cheap way of joining a demand management scheme which pays a company to reduce its consumption during peak demand periods.

Alternative fuels are another area where microturbines are beginning to find a niche market. Methane gas from small landfill sites offers one possibility but so does biogas generated from a variety of sources. The simplicity of a packaged microturbine may make it more attractive than a reciprocating gas engine, the type of engine commonly used to burn these gases, in situations where a relatively small amount of gas is available.

Where there is a need for heat as well as electricity, microturbines can offer a cost-effective source of energy. If the site is in an urban area where emissions regulations are tight, the microturbine could be the easiest solution to install. In the United States there are examples of microturbine installations in restaurants and on a university campus.

8.4 ULTRA-MICRO GAS TURBINES

There is a type of gas turbine called an ultra-microturbine that is even smaller than the micro gas turbine. These ultra-micro gas turbines are still in the research stage of their development but they offer intriguing possibilities. The machines offer outputs of tens to hundreds of watts and are typically only a few centimetres in diameter. With machines this tiny, fabrication of parts such as the turbine and compressor present major problems and techniques such as those used to manufacture microchips are being exploited. These tiny turbines can rotate at more than 100,000 rev/minutes. The small size means that a variety of effects that are not significant in large machines, such as air flow friction, will play a major role in their operation. Novel designs for combustion chambers are also required and new materials may be needed to build all the components.

Efficiency of these tiny gas turbines is expected to be lower than for the microturbines discussed above. Their advantage is power density, particularly when compared to batteries which are their likely competitors. Applications for such devices include use by military personnel and in drones, the latter offering perhaps the most attractive market. Civilian uses may be limited because of the high operating temperature of the devices.

CHAPTER *9*

Gas-Fired Power Plants and the Environment

Natural gas is a fossil fuel and its combustion has a significant environmental impact. Combustion releases large quantities of carbon dioxide into the atmosphere and this contributes to the increase in atmospheric concentrations of the gas which are blamed for global warming. Methane, the major constituent of natural gas as supplied by pipeline, is on its own a potent greenhouse gas so any inadvertent release of this will also be important. According to the US Environmental Protection Agency, 3.8% of greenhouse gas emissions in the United States between 1990 and 2009 came from methane released by oil and natural gas systems. Most of this is released during oil and gas production but transmission and distribution accounts for around 30% of the annual total.

In addition to these greenhouse gas emissions associated with natural gas-fired power plants, there are damaging noxious emissions from the combustion process itself. The most important of these for a gas plant is the generation of nitrogen oxides from the nitrogen in air at the high combustion temperatures typical of natural gas combustion. These emissions have been linked to a range of environmental problems including a serious impact on human health. Depending on the combustion conditions, combustion of natural gas can also generate carbon monoxide as well as some unburnt hydrocarbons. Both must be controlled to meet modern emission regulations.

Unlike coal, natural gas contains very little sulfur and so this has no significant environmental impact. It contains no heavy metals either. However, it will release waste heat to the environment both in the form of hot exhaust gases through the plant stack and warm water if local water supplies are used in the steam turbine condenser of a combined cycle plant. Heat emissions of this type are not usually considered particularly harmful but they can change local environmental conditions.

Gas-Turbine Power Generation. DOI: http://dx.doi.org/10.1016/B978-0-12-804005-8.00009-4

Like any other power plant, there will be some local disruption due to additional traffic movements associated with the plant. The fuel, gas, will be delivered by pipeline so this will cause little disruption once it has been built. Overall, a natural gas-fired power station is likely to be the most benign of any fossil-fuel powered plant.

9.1 NITROGEN OXIDE EMISSIONS

The combustion of fossil fuels including natural gas leads to the production of quantities of nitrogen oxides, often collectively referred to as NO_x. The main constituents of NO_x are nitrogen oxide (or nitric oxide, NO), nitrogen dioxide (NO_2), and nitrous oxide (N_2O). In power plant combustion furnaces the main NO_x constituent generated is nitrogen oxide which is the most stable oxide at these temperatures. It probably accounts for 90% or more of the NO_x from a gas-fired plant.

Nitrogen oxides are controlled because of the range of environmental problems they can cause. They are a major source of ground level ozone which can cause severe respiratory problems. The gas can react with other components in the atmosphere to create an acid aerosol which is also harmful. NO_x is responsible in part for the haze that falls over urban areas and it can, together with sulfur dioxide, cause acid rain which has been responsible for destroying forests and life in lakes as well as damaging buildings. An abundance of nitrogen from nitrogen oxides is a cause of massive algae growths in waterways. Finally, nitrous oxide is a greenhouse gas.

Table 9.1[1] shows a selection of national and regional standards for the emission of nitrogen oxides from power plants. In all cases the

Table 9.1 International Emission Standards for Nitrogen Oxides from Power Plants		
Country/Region	Emission Limit for New Plants (mg/m³)	Emission Limit for Existing Plants (mg/m³)
China	100	160–640 for plants built before 2006
European Union	200	500 for plants built before 2016
United States	117	200 for plants built before 2004
Source: *AirClim*.		

[1]China: New emission standards for power plants, Acid News, AirClim, October 2012.

standards have been tightened in recent years as the dangers of NO_x emissions have become clearer. In China all new plants have to adhere to a limit of $100\,mg/m^3$. For existing plants, those built after 2005 also have to meet this standard but older plants have less stringent limits. In the European Union the limit is $200\,mg/m^3$ for all plants after 2016 but older plants can emit $500\,mg/m^3$ before that date, although they must comply with the new limit from the start of 2016. In the United States the standard is $117\,mg/m^3$ for plants built after 2005 but earlier plants can release up to $640\,mg/m^3$. All gas-fired combined cycle power plants within these jurisdictions must comply with these limits. Similar limits apply elsewhere.

There are two principle strategies to control the NO_x emissions from a gas-fired power plant. The first is by careful control of the combustion conditions when the natural gas is burnt and the second is by removal of the NO_x from the exhaust gas stream of the power plant after it has been produced. Most natural gas-fired power plants need to use both strategies to comply with emissions regulations.

9.2 NO_x PRODUCTION PATHWAYS

Natural gas contains no significant amount of nitrogen, unlike coal, and so all the NO_x produced in a gas-fired power plant is the consequence of the oxidation of nitrogen from air at the very high temperatures generated during the combustion process. There are probably hundreds of different chemical pathways that can lead to NO_x formation, many involving different fragmentary reactants that are generated during the combustion.

Among this multitude of possibilities, two broad pathways have been identified, one called thermal generation and the other prompt generation. Thermal generation involves a reaction during the combustion process which releases oxygen atoms from oxygen molecules in air. The freed oxygen atoms then react with molecular nitrogen to generate nitrogen oxide. This reaction is important once the combustion temperature rises above around 1300°C and will become the major source of nitrogen oxide above 1600°C when both molecular nitrogen and molecular oxygen start to dissociate spontaneously.

The prompt pathway—which is a catch-all for the other ways NO_x can be produced—normally involves molecular fragments called hydrocarbon radicals generated from the breakup of hydrocarbon molecules. These react with molecular nitrogen producing a range of nitrogen-containing radicals that are easily converted to nitrogen oxide. As with thermal production of nitrogen oxides, the higher the temperature the more likely they are to be produced.

Since the speed of all these reactions depends on the combustion temperature, so does the quantity of NO_x that is produced during combustion. The higher the temperature, the greater the quantity produced. At the same time the efficiency of a heat engine depends on temperature; high efficiency requires high working gas temperatures. A high efficiency engine must therefore lead to conditions that are ideal for the production of large quantities of NO_x.

9.3 LOW NO$_X$ BURNERS

In spite of these competing requirements, it is possible to limit the amount of NO_x produced during the high temperature combustion process. The most important strategy is to control the amount of oxygen that is available in the hottest part of the combustion flame.

Methane (CH_4) and its constituents, hydrogen and carbon, are all more reactive towards oxygen than nitrogen. Each molecule of methane requires two molecules of oxygen for complete combustion. If the amount of oxygen is limited to this amount or less, methane molecules will preferentially react with the oxygen that is available, limiting the amount that of oxygen can be turned into NO_x. If the amount of oxygen is less than is required[2] the conditions will lead to incomplete combustion of the natural gas. The reaction must eventually be allowed to go to completion if all the energy from methane combustion is to be released so more air is added to the combustion chamber later, when the combustion gases have cooled a certain amount and the likelihood of NO_x production is reduced.

In a gas-fired steam plant, this strategy is called staged combustion. Air and natural gas are mixed in less than stoichiometric proportions and then ignited in the boiler to create the main fireball. Further air is then admitted to the hot combustion gases in the region above the fireball where the temperature is lower. It is also possible to admit

---

[2]This is often referred to as reducing conditions.

additional natural gas above the main combustion zone to preferentially scavenge oxygen from nitrogen oxides, reducing them to nitrogen again. In order for the combustion reaction to continue fully it is normal to add an excess amount of air above that needed for stoichiometry. The excess air also helps cool the combustion gases. For natural gas, the air to natural gas ratio, by volume, required for a stoichiometric amount of oxygen to be present is 1:8.43.

The situation is slightly more complex in a gas turbine combustor because the total amount of air that is mixed with the natural gas has to be controlled in order to keep parasitic losses resulting from the compression of excess air to a minimum. Nevertheless a similar type of staged combustion can be achieved, with air and fuel mixed in less than stoichiometric proportions for initial combustion while additional air is admitted through the combustor liner at later stages. The method used in most advanced gas turbine combustors is slightly different, a process called premixed combustion in which air and natural gas are carefully and thoroughly mixed before entering the combustor with just enough oxygen present to react with the combustible constituents of the natural gas and none left for nitrogen production. As with staged combustion, this process must be carefully and continuously controlled. Any changes in natural gas composition must be monitored so that the amount of air can be adjusted as necessary. Too little oxygen will lead to production of carbon monoxide and unburnt hydrocarbon fragments which are potentially harmful. Too much oxygen and NO_x levels will rise. With complex fuel nozzle configurations and careful air/fuel mixing the production of NO_x can be limited to 15–25 ppm for large industrial gas turbines and under 10 ppm in some small turbines designs.

The injection of water or steam into the combustion chamber has also been used to limit the production of NO_x by cooling the combustion gases so that the reaction rate for nitrogen compounds is reduced. This can compensate for limited amounts of excess air. Water/steam injection also has an impact on the overall cycle performance. It is not normally used in large industrial turbines but is used in some aero derivative gas turbines.

9.4 SELECTIVE CATALYTIC REDUCTION

While controlled combustion can limit the production of NO_x it is unlikely to reduce the levels low enough to meet emissions regulations.

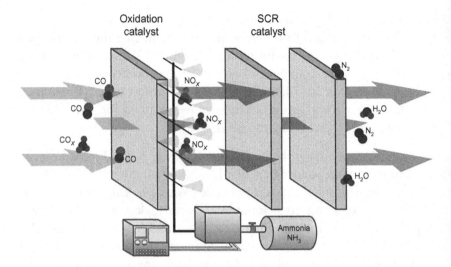

Figure 9.1 Schematic of a catalytic oxidation and SCR system for a gas turbine power plant. Source: From Rentech Boilers.

Therefore an additional procedure is required. The most commonly used is selective catalytic reduction (SCR). This process involves mixing the exhaust gases with a suitable reducing reagent and then passing them over a specially formulated metallic catalyst. The catalyst promotes the reaction between the reducing reagent and nitrogen oxides in the exhaust gas, turning them back into molecular nitrogen.

The reducing agents used in the catalytic reduction process are either ammonia, which is the most common reagent, or urea. The reagent must be mixed with the exhaust gases at a temperature that is suitable for the reaction to proceed. For the SCR process[3] this is typically between 250°C and 430°C but can fluctuate by around 90°C either side of this range and still be effective. The layout of an SCR system is shown in Fig. 9.1.

The other key ingredient is the catalyst. This is usually an active metal or ceramic with a highly porous structure. A typical catalyst for SCR might contain around 80% titanium dioxide as the main substrate with vanadium pentoxide, tungsten trioxide, and a small amount of aluminum. However, most catalysts are proprietary and compositions

[3]The reaction can be carried out without a catalyst but it requires a much higher temperature and is generally not so efficient at removing NO_x. This process, called selective noncatalytic reduction is sometimes used in coal-fired power stations but rarely in gas turbines.

vary. The catalyst normally has a honeycomb structure which presents a large surface area to the exhaust gases which pass through the catalytic modules. The catalyzed reaction takes place on the surface of the catalyst so the larger the area available, the more effective it is.

In order for the exhaust gases and reducing reagent to enter the catalyst module at the correct temperature, the SCR system is often installed within the heat recovery steam generator of a combined cycle plant. For an open cycle gas turbine system it may be necessary to cool the exhaust gases before they enter the catalytic system. Once the flue gases are cool enough, they are mixed with ammonia from a spray system in the flue gas ductwork. The catalytic modules which catalyze the reaction are placed downstream, where the NO_x molecules are converted into nitrogen and water vapor. These harmless products are then carried away in the flue gases and exhausted to the atmosphere.

The overall efficiency of the SCR process depends on the concentration of NO_x in the exhaust gases. At around 20 ppm the efficiency is close to 70% but may fall at lower concentrations. This efficiency will rise with concentration up to around 150 ppm. High efficiency SCR systems can achieve greater than 90% conversion efficiency.

It is important to control the amount of ammonia that is added to the exhaust gases during SCR because if there is too much, some will remain in the flue gases that are released into the environment, creating a new emissions problem. To control this "ammonia slip" requires monitoring of both the quantity of NO_x in the exhaust gases and the ammonia that remains in the flue gases after the SCR reactor. If sulfur is present, the SCR reactor can turn it into sulfur trioxide which will react with water to produce highly corrosive sulfuric acid. This is generally not a problem with natural gas but could be in a dual-fuel gas turbine plant if it switches to a liquid fuel that contains any sulfur.

The SCR catalyst is sensitive to certain flue gas constituents that can poison it, reducing its effectiveness. However, natural gas contains few, if any, constituents that will harm the catalyst and this techniques is successfully and economically applied to most large gas turbine and combined cycle plants. Too high a temperature can also impair its effectiveness so the flue gas temperature must be controlled. When the efficacy of the catalyst modules falls below a certain level they must be removed and replaced. The exhausted modules will generally be regenerated or their catalytic constituents recycled for reuse.

9.5 CARBON MONOXIDE

Natural gas combustion in a gas turbine can lead to high levels of carbon monoxide during controlled combustion. If there is insufficient oxygen present then the combustion reaction does not run to completion and CO is formed instead of CO_2. Carbon monoxide production can also increase when the gas turbine is operating at part load because the temperature in the combustion chamber is lower, again leading to incomplete combustion. At the same time, incomplete combustion can also produce unburnt hydrocarbon fragments known as volatile organic components or VOCs.

Combustor design can help promote full combustion but that may not be sufficient to keep the production of CO below regulatory emission limits. If that is the case then a carbon monoxide control system must be installed. This is normally a catalytic oxidation system. As with the SCR system described above, a catalytic oxidation system has catalytic modules that are placed in the exhaust gas path. These contain metallic catalysts that catalyze the reaction of carbon monoxide with some of the oxygen remaining in the flue gases, converting it to carbon dioxide. The same catalyst will usually be effective in oxidizing VOCs too, rendering them harmless.

In order to prevent interaction between different catalytic systems a catalytic oxidation system will usually be placed before the SCR system, as shown in Fig. 9.1. Oxidation catalysts include metals like rhodium, platinum, palladium, ruthenium and iridium. As with SCR catalytic modules, these metals will be carried in a ceramic honeycomb that presents a very large surface area to encourage rapid reaction of the carbon monoxide and VOCs as the pass through the modules.

9.6 CARBON DIOXIDE

Carbon dioxide is one of the normal combustion products when a hydrocarbon such as methane is burnt in air or oxygen. The basic combustion reaction of methane in air is:

$$CH_4 + 2O_2 = CO_2 + 2H_2O$$

While methane does not produce as much carbon dioxide as coal does for each unit of electricity produced, the greenhouse gas is still a major product that is released into the atmosphere from all natural gas-fired power plants. In the United States in 2014 natural gas-fired power plants accounted for 22% of power sector carbon dioxide emissions and around 27% of generated power according to the US Energy Information Administration. While the shift to natural gas has not been as pronounced in most other regions as it has in the United States, emissions will still be considerable.

The control of carbon dioxide emissions from power stations is not mandated anywhere although restrictions are under discussion.[4] In consequence there are as yet no commercial systems that will remove carbon dioxide from the flue gases of a natural gas-fired power plant. However, there are ways in which these emissions could be controlled and these may well be required by the middle of the third decade of the 21st century.

There are three ways of reducing the carbon dioxide emissions from a natural gas-fired power plant, postcombustion capture, precombustion capture and oxyfuel combustion. Application of any of them will reduce the efficiency of a gas-fired combined cycle power plant by 7 to 17 percentage points compared to a plant with no carbon capture. (The discussion of the three processes below is primarily concerned with gas turbine combined cycle plants since these represent the bulk of natural gas-fired power plants. However, similar techniques can be applied to natural gas-fired steam plants.)

Postcombustion capture is conceptually the simplest scheme to apply. It involves fitting a chemical processing plant to the power station, a processing plant that scrubs the exhaust gases to remove carbon dioxide. This technique is already used in the oil and gas industry and the technology for capture is well understood. One of its major advantages is that it can be retrofitted to existing plants. However it has never been applied to a commercial combined cycle power plant.

The process involves passing the flue gases up through a tall tower into the body of which is sprayed a reagent that will selectively react with and remove the carbon dioxide from the gases. The concentration of carbon dioxide in the flue gases of a combined cycle plant are typically 3–4%, relatively low when considering the reaction mechanics of

---

[4]There are systems such as the European Union Emissions Trading System that limit total emissions from certain sectors and charge for those emissions. Depending on the cost for the emission of a ton of carbon dioxide, systems like this can act to control emissions.

the capture process. In consequence this reagent needs to form a chemical bond with the carbon dioxide to be effective. Once the reagent has captured the carbon dioxide, it is cycled through a regeneration plant where it is heated to release the carbon dioxide. The latter is then compressed ready for transportation and storage, while the reagent is ready for reuse. This regeneration process is energy intensive because of the chemical bond between the CO_2 and the scrubbing reagent and regeneration is responsible for the larger part of the energy losses resulting from carbon capture. The most common capture reagent is monoethanolamine but other reagents are being developed. The efficiency penalty for postcombustion capture is around 7–8 percentage points.[5]

Precombustion capture is essentially chemical reforming of natural gas to convert it into hydrogen, followed by capture of the carbon dioxide produced during the reforming process. Reforming is a partial oxidation process and it normally requires an oxygen plant when carbon capture is being applied because otherwise the gases resulting from the reaction will be diluted with nitrogen from air, making the separation of carbon dioxide more difficult. When an oxygen plant is used, the final product is virtually pure hydrogen which is burnt in a gas turbine combined cycle plant. The efficiency penalty for this process is much higher than for postcombustion capture, with recent International Energy Agency estimates putting it at 16–22 percentage points.

The final scheme, oxyfuel combustion, replaces the combustion air in the gas turbine with oxygen. This means that the product of combustion is mainly carbon dioxide and water vapor, allowing the carbon dioxide to be separated easily. However, it too requires an oxygen plant. In effect oxyfuel combustion replaces the separation of carbon dioxide from nitrogen in the exhaust gases of a plant for the separation of oxygen from nitrogen in air before combustion. In addition the use of oxygen instead of air changes the operating conditions of the gas turbine both by significantly increasing the combustion temperature compared to combustion in air and by changing the mass flow through the turbine. This complicates the design of a combined cycle power plant with oxyfuel combustion. Oxyfuel combustion is seen as attractive in coal-fired power stations, and may therefore also be attractive in natural gas steam plants but it is probably less cost-effective than postcombustion capture in a combined cycle plant.

---

[5]CO_2 capture at gas-fired power plants, IEA, July 2012.

CHAPTER 10

The Cost of Electricity Generation from Natural Gas-Fired Power Plants

The cost of electricity from a power plant of any type depends on a range of factors. First there is the cost of building the power station and buying all the components needed in its construction. In addition, most large power projects today are financed using loans so there will also be a cost associated with paying back the loan, with interest. Then there is the cost of operating and maintaining the plant over its lifetime. Finally the overall cost equation should include the cost of decommissioning the power station once it is removed from service.

It would be possible to add up all these cost elements to provide a total cost of building and running the power station over its lifetime, including the cost of decommissioning, and then dividing this total by the total number of units of electricity that the power station produced over its lifetime. The result would be the real lifetime cost of electricity from the plant. Unfortunately, such a calculation could only be completed once the power station was no longer in service. From a practical point of view, this would not be of much use. The point in time at which the cost-of-electricity calculation of this type is most needed is before the power station is built. This is when a decision is made to build a particular type of power plant, based normally on the technology that will offer the least cost electricity over its lifetime.

10.1 LEVELIZED COST OF ENERGY MODEL

In order to get around this problem economists have devised a model that provides an estimate of the lifetime cost of electricity before the station is built. Of course, since the plant does not yet exist, the model requires a large number of assumptions and estimates to be made. In order to make this model as useful as possible, all future costs are also converted to the equivalent cost today by using a parameter known as the discount rate. The discount rate is almost the same as the interest

Gas-Turbine Power Generation. DOI: http://dx.doi.org/10.1016/B978-0-12-804005-8.00010-0

rate and relates to the way in which the value of one unit of currency falls (most usually, but it could rise) in the future. This allows, for example, the maintenance cost of a gas turbine 20 years into the future to be converted into an equivalent cost today. The discount rate can also be applied to the cost of electricity from the wind power plant in 20 years time.

The economic model is called the levelized cost of electricity (LCOE) model. It contains a lot of assumptions and flaws but it is the most commonly used method available for estimating the cost of electricity from a new power plant.

When considering the economics of new power plants the levelized cost is one factor to consider. Another is the overall capital cost of building the generating facility. This has a significant effect on the cost of electricity but it is also important because it shows the financial investment that will have to be made before the power plant generates any electricity. The comparative size of the investment needed to build different types of power stations may determine the actual type of plant built, even before the cost of electricity is taken into account. This is of particular significance with gas turbine combined cycle plants as their capital cost is among the lowest of all generating technologies. Capital cost is usually expressed in terms of the cost per kilowatt of generating capacity to allow comparisons between technologies to me made.

When comparing different types of power station there are other factors that need to be considered too. The type of fuel, if any, that it uses is one. A coal-fired power station costs much more to build than a gas-fired power station but the fuel it burns is relatively cheap. Natural gas is more expensive than coal and it has historically shown much greater price volatility than coal. This means that while the gas-fired station may require lower initial investment, it might prove more expensive to operate in the future if gas prices rise dramatically.

Renewable power plants can also be relatively expensive to build. However, they normally have no fuel costs because the energy they exploit is from a river, from the wind, or from the sun and there is no economic cost for taking that energy. That means that once the renewable power plant has been paid for, the electricity it produces will have a very low cost. All these factors may need to be balanced when making a decision to build a new power station.

Table 10.1 Overnight Capital Cost of Gas Turbine and Combined Cycle Plants			
Year	Advanced Combined Cycle Plant ($/kW)	Advanced Combined Cycle Plant With Carbon Sequestration ($/kW)	Advanced Open Cycle Gas Turbine ($/kW)
2001	533	–	440
2003	563	–	439
2005	517	992	356
2007	550	1055	379
2009	877	1683	604
2011	917	1813	626
2013	931	1833	632
2015	942	1845	639
Source: US Energy Information Administration.			

10.2 CAPITAL COST

The capital cost of a natural gas-fired power plant depends on the type of plant. A gas-fired steam raising plant will have costs that are broadly comparable to a coal-fired power plant, although there will be savings on the fuel handling side and the gas-fired plant will not need either a sulfur scrubber or an dust removal unit. The boiler and steam turbine will be the dominant cost items.

The capital cost of gas turbine power plants is determined primarily by the cost of the turbine generator. These are high technology machines and the number of manufacturers is limited but the market is global and competition is fierce. In addition, these are essentially off-the-shelf components that are delivered from a factory virtually ready to install and operate. This means that the time required to build a gas turbine power plant is much shorter than, for example, a coal-fired power plant. These factors have led to gas turbine combined cycle plants becoming the cheapest large capacity power plants to build.

Table 10.1[1] shows some typical capital cost figures. These are from the US Energy Information Administration (US EIA) and show figures from the organizations Annual Energy Outlooks from 2001 to 2015.[2] The cost of a gas turbine depends on the cost of a range of high technology materials and these fluctuate on the global commodities market. Even so,

[1]Data are taken from the US Energy Information Administration Assumptions to the Annual Energy Outlook 2001–2015.
[2]The capital cost figures in table are for the year before the report year shown in column one of the table and are calculated in $/kW from the preceding year.

excepting a large jump in costs between 2007 and 2009, when commodity costs peaked, the price changes have been relatively steady.

In 2001 the capital cost of an advanced combined cycle plant was $533/kW. The overnight cost rose slowly to $563/kW in 2003 before falling back to $550/kW in 2007. In 2009 the cost was estimated to be $877/kW and has risen steadily since, reaching $942/kW in 2015. The cost of an advanced open cycle gas turbine has shown a similar trend, with a cost of $440/kW in 2001, rising to $639/kW in 2015. The table also contains estimated costs for an advanced combined cycle plant with carbon sequestration, a configuration that might become mandatory over the next decade. The cost of this plant was put at $992/kW in 2005, 92% higher than a plant without carbon sequestration. In 2015 the estimated cost of this plant was $1845/kW, 96% more expensive than the plant without carbon capture. By way of comparison, US EIA Annual Energy Outlook for 2015 estimates cost of a coal-fired power plant to be $2726/kW and that of an onshore wind farm at $1850/kW.

These figures are all for the United States. Elsewhere in the world the capital cost varies, but not widely. According to the International Energy Agency (IEA), the capital cost of a combined cycle gas turbine plant within the Organization for Economic Co-operation and Development (OECD) ranges from $45/kW in Korea to $1289/kW in New Zealand.[3]

Microturbine capital costs are similar to those of their larger relatives. The cost of a microturbine without heat recovery is estimated to be $700/kW to $1100/kW. A microturbine cogeneration plant with heat recovery will have costs around 30% to 50% higher.[4]

10.3 FUEL COSTS

Combined cycle power plants are relatively cheap to build. However, the cost of the electricity they produce depends critically on the cost of natural gas. This has shown a historical tendency to fluctuate. When fuel costs are low the cost of electricity from a natural gas-fired plant will be low. However, when the cost rises, it can make gas-fired power plants uneconomical to operate and there are examples in many developed countries of combined cycle plants being shut down because they cannot generate electricity competitively.

[3]Projected Costs of Generating Electricity, 2015 Edition, IEA/NEA, 2015.
[4]Figures are from the California Distributed Energy Resources Guide on Microturbines.

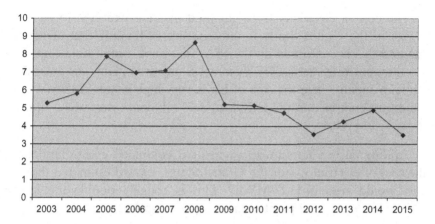

Figure 10.1 Average annual cost of natural gas to US utilities. Source: With permissiomn fom Power Research Institute, Inc. US Energy Information Electric Power Monthly October 2015.

The cost of gas depends on availability and in the United States, the advent of shale gas production has led to a fall in gas prices, making gas-fired generation extremely attractive. Fig. 10.1 shows the annual average cost of natural gas to US electric utilities between 2003 and 2015. The cost was between $5/GJ and $6/GJ at the beginning of the period and peaked at close to $9/GJ in 2008. However, by 2015 the cost had fallen to under $4/GJ.

In other parts of the world prices are generally much higher. In 2014 the average cost of natural gas in the United Kingdom was just under twice that in the United States. In Japan, meanwhile the cost of gas including the transportation costs was almost four times the cost in the United States. In these regions, the cost of gas-fired electricity is much higher than in the United States.

10.4 THE LCOE FROM A NATURAL GAS-FIRED POWER STATION

The LCOE from a natural gas-fired combined cycle power plant in the United States in 2014 was estimated by Lazard to be between $61/MWh and $87/MWh.[5] Only wind power was estimated to be consistently cheaper.

---

[5]Lazard's Levelized Cost of Energy Analysis—Version 8.0, 2014.

The cost of electricity from gas-fired plants is expected to be at its lowest in the United States in consequence of the low cost of natural gas and this is confirmed by the IEA. Its recent study found that, at a 7% discount rate, the LCOE from a US gas-fired plant was $66/MWh while in Japan where the cost of gas is very high it was $138/MWh. In China, in comparison the LCOE at a 7% discount rate was estimated to be $93/MWh.[6]

Adding carbon capture to a gas-fired combined cycle plant raises the cost of electricity. For the United States, Lazard estimated that the cost would rise to $127/MWh.[7] This is roughly double the lowest LCOE for a gas-fired plant without carbon capture.

---

[6]Projected Costs of Generating Electricity, 2015 Edition, IEA/NEA, 2015.
[7]This price does not include the cost of transportation and storage of the captured carbon dioxide.

Printed in the United States
By Bookmasters